Charles Loudon Bloxam

Laboratory Teaching

Or, progressive exercises in practical chemistry

Charles Loudon Bloxam

Laboratory Teaching
Or, progressive exercises in practical chemistry

ISBN/EAN: 9783337164157

Printed in Europe, USA, Canada, Australia, Japan

Cover: Foto ©Paul-Georg Meister /pixelio.de

More available books at **www.hansebooks.com**

LABORATORY TEACHING:

OR,

PROGRESSIVE EXERCISES

IN

PRACTICAL CHEMISTRY.

BY

CHARLES LOUDON BLOXAM,
PROFESSOR OF CHEMISTRY IN KING'S COLLEGE, LONDON;
IN THE DEPARTMENT OF ARTILLERY STUDIES, WOOLWICH, AND IN THE
ROYAL MILITARY ACADEMY, WOOLWICH.

FOURTH EDITION,

WITH

EIGHTY-NINE ILLUSTRATIONS.

PHILADELPHIA:
LINDSAY AND BLAKISTON.

PREFACE

TO

THE FOURTH EDITION.

The most important alteration in the present Edition is the introduction of the formulæ representing the various chemical compounds described in the Notes to the Tables. The formulæ are those now generally employed by chemical writers and teachers in this country.

The verbal description of the composition in the Tables of Common Compounds of the several metals has not been altered so as to bring it into perfect harmony with the formulæ, since the description there given generally informs the learner what substances can be obtained by the decomposition of the Common Compounds, which is not so easily to be ascertained by an inspection of the formulæ.

For example, the composition of saltpetre is described at page 81, as Potash (Potassium and Oxygen) and Nitric Acid, whilst the formula KNO_3 does not indicate the presence of potash (K_2O) or of nitric acid (HNO_3); but both these substances are obtainable from saltpetre by very simple chemical operations, and saltpetre may be produced by causing them to act upon each other.

It is true that similar reasoning would justify the statement that common salt contained soda and hydrochloric acid instead of sodium and chlorine, but the Author feels that an endeavor to be absolutely consistent would injure the practical usefulness of so small a book.

May, 1879.

PREFACE

TO

THE FIRST EDITION.

This work is intended for use in the Chemical Laboratory by those who are commencing the study of Practical Chemistry. It contains—

(1) A series of simple Tables for the analysis of unknown substances of all kinds (not excepting organic bodies) which are known to be single substances, and not mixtures:

(2) A brief description of all the practically important single substances likely to be met with in ordinary analysis, by which the learner may satisfy himself that his results are correct, and may at the same time become acquainted with the leading properties of the most important chemicals, and with the foreign substances which they are liable to contain:

(3) Simple directions and illustrations relating to Chemical Manipulation, not collected into a separate chapter, but given just where the learner requires them in the course of analysis:

(4) A system of Tables for the detection of unknown substances with the aid of the Blowpipe:

(5) Short instructions upon the purchase and preparation of the tests, intended for those who have not access to a Laboratory.

The book does not presuppose any knowledge of Chemistry on the part of the pupil, and does not enter into any theoretical speculations.

It dispenses with the use of all costly Apparatus and Chemicals, and is divided into separate Exercises or Lessons, with Examples for Practice, to facilitate the instruction of large Classes.

The Author hopes that it will be found to contain all the Practical Chemistry required for the various Examinations, except for the highest Science degrees, such as the B.Sc. and D.Sc. of the University of London.

The method of instruction here followed has been adopted by the Author after twenty-three years' experience as a teacher in the Laboratory, by which he has been led to conclude that a knowledge of Analytical Chemistry, or the power of discovering the nature of unknown substances, is the first and often the only requirement of the great majority of learners, and that, independently of the technical value of such knowledge, its acquisition forms a most valuable part of education, by cultivating the powers of observation, and affording excellent examples of the application of logical reasoning in practical work.

The ordinary method of teaching Analytical Chemistry, by causing the pupil to study the *Reactions* or tests for all the metals and non-metallic bodies, before he proceeds to exa-

PREFACE TO THE FIRST EDITION.

mine an unknown substance, does, without doubt, lay the best foundation for a thorough knowledge of this branch of Science, when a student has time at his command, and can commit to memory a large number of independent facts which receive no immediate application. But, within the Author's experience, many students become wearied with the monotonous routine of this system, and are prevented from persevering in the study.

Moreover, such a system, although teaching the student to discover, for example, that a given salt contains potassium and nitric acid, fails often to instruct him that these constitute saltpetre, and does not acquaint him with the appearance and other properties of saltpetre, by observing which he may be sure that his analysis is correct. For want of such an acquaintance with the properties of the common salts, a student well skilled in the mere detection of bases and acids will sometimes fall into the most absurd mistakes as to the nature of a substance under examination.

Those students who can afford the time are strongly recommended to perform every experiment described in the book, with the known substances, before attempting the examination of an unknown substance.

April, 1869.

*** *A List of the Apparatus and Chemicals required for the Course will be found at pp.* 228–243.

CONTENTS.

PARAGRAPHS 1—31.—Analysis of substances containing a single metal. Solution. Filtration. Precipitation. Division of the metals into groups. Identification of the individual members of each group. Description of the metals and of their chief oxides and salts.

PARAGRAPHS 82—121.—Analysis of substances containing a single inorganic acid or non-metallic body. Expulsion of acids from their compounds by sulphuric and hydrochloric acids. Precipitation of acids by barium nitrate, silver nitrate, calcium chloride and ferric chloride. Description of the principal inorganic acids.

PARAGRAPHS 122—141.—Analysis of substances insoluble in water and acids. Fusion. Description of the principal insoluble substances.

PARAGRAPHS 142—162.—Analysis of substances which may contain one metal and one organic acid. Description of the principal organic acids, and their common salt.

PARAGRAPHS 163—171.—Detection of the principal vegetable alkaloids. Identification of caffeine, morphine, brucine, strychnine, quinine, narcotine, and cinchonine.

PARAGRAPHS 172—176a.—Identification of the more common organic substances characterized by color or odor. Indigo. Picric acid. Caramel. Carbolic acid. Chloral hydrate.

PARAGRAPHS 177—205.—Examination of a solid organic substance having no definite color or odor. Identification of cane-sugar, grape-sugar, milk-sugar, urea, pyrogalline, salicine, albumen, starch, dextrine, gum, gelatine, soap, stearine, stearic acid, palmitic acid, cholesterine, rosin, naphthaline, palmitine, spermaceti, wax, and paraffine.

PARAGRAPHS 206—229.—Examination of a liquid organic substance having a very distinct odor. Identification of alcohol, methylated spirit, wood-naphtha, acetone, aldehyde, nicotine, butyric acid, valerianic acid, aniline, ether, chloroform, oil of bitter almonds, nitrobenzole, and benzole. Distillation.

PARAGRAPHS 230—234.—Examination of a liquid organic substance having no distinct odor. Identification of glycerine, lactic acid, oleine, and oleic acid.

PARAGRAPH 235.—Examination of a solid organic substance about which nothing is known but that it is a single substance and not a mixture.

PARAGRAPH 236.—Examination of a liquid organic substance about which nothing is known but that it is a single substance and not a mixture.

PARAGRAPH 237.—Examination of a solid substance of which nothing is known but that it is a single substance and not a mixture.

PARAGRAPH 238.—Examination of a liquid of which nothing is known but that it is a solution of a single substance and not a mixture.

PARAGRAPH 239.—Examination of an organic substance which is known to be included in the list given for the 1st M.B. Examination of the University of London.

PARAGRAPHS 240—282.—Detection of metals by the blowpipe. Reduction on charcoal. Borax-Beads. Colored flame test. Cobalt test. Sublimation test. Cupellation.

PARAGRAPHS 283—307.—Detection of non-metals or acids by the blowpipe. Heating with bisulphate of potash.

PARAGRAPHS 308—363.—Alphabetical list of test used in analysis.

PARAGRAPH 364.—Apparatus required in qualitative analysis.

APPENDIX.—Qualitative Analysis of Gunpowder.

LABORATORY TEACHING;

OR,

PROGRESSIVE EXERCISES IN PRACTICAL CHEMISTRY.

INTRODUCTION.

The instructions given in this work are intended to enable the student to discover the nature of most of the chemical compounds in common use.

When nothing is known about the substance, its examination must be commenced according to Exercise XIII. (paragraph 237) if it be a solid, or XIV. (paragraph 238) if it be a liquid.

But it is better for beginners to confine their attention to ordinary inorganic salts, which may be commenced by Table A. For the first few lessons, the detection of the metal, by Tables A to F, will suffice.

Having detected the metal, the student should turn to the description of that metal and its common compounds, and endeavor to name the unknown body under examination.

When he is well versed in the detection of the metal, he may proceed to discover the non-metallic body or acid which composes the other portion of the compound.

In many cases, where a substance has a peculiar color or smell, a reference to that color or smell in the Index may lead to its identification.

The learner will find it advantageous to study the list of the principal tests, commencing with paragraph 308, since he will there find the formulæ expressing their composition, and a list of the substances, the presence of which is indicated with certainty by the action of any particular test.

The numerals inclosed in parentheses in each Table refer to paragraphs in the text, and it will be found absolutely necessary to refer to these in order to render the Tables serviceable.

A rough note-book should be provided, in which every step of the analysis should be entered, the symbols being employed for the tests, both to avoid much writing and to fix their composition in the memory.

A Report of an Analysis is appended as a model.

REPORT OF ANALYSIS.

No. 12.

Colorless Crystals. Dissolved in H_2O, Alkaline.*

NH_4Cl NH_3 $(NH_4)_2S$ 0	NH_4Cl NH_3 $(NH_4)_2CO_3$ 0	KHO Boiled 0	$H_2C_4H_4O_6$ stirred on glass 0	$PtCl_4$ stirred on glass 0	$KSbO_3$ stirred on glass Ppt.	Flame, golden yellow
HCl 0 H_2S 0 Boiled 0	Boiled 0 Na_2HPO_4 Stirred 0					Metal is Na.

* See paragraph 18.

For Acid.

HCl 0 Heated 0	$Ba(NO_3)_2$ white ppt. HNO_3 dissolved	$AgNO_3$ white ppt. HNO_3 dissolved	$CaCl_2$ white ppt. $HC_2H_3O_2$ dissolved Fe_2Cl_6 0	Fe_2Cl_6 brown ppt. $HC_2H_3O_2$ dissolved	HCl and turmeric paper; dried; pink; KHO green
H_2SO_4 0 Heated 0					Acid is Boracic.

Fused on Pt wire to transparent bead.
Substance is borax.

TABLE OF ATOMIC WEIGHTS.

The subjoined Table indicates the relative weights of the Elementary Bodies represented by the symbols composing the formulæ given in this work.

Aluminium	Al	27·5	Lead	Pb	207
Antimony	Sb	122	Magnesium	Mg	24·3
Arsenic	As	75	Manganese	Mn	55
Barium	Ba	137	Mercury	Hg	200
Bismuth	Bi	210	Nickel	Ni	59
Boron	B	10·9	Nitrogen	N	14
Bromine	Br	80	Oxygen	O	16
Calcium	Ca	40	Phosphorus	P	31
Carbon	C	12	Platinum	Pt	197·1
Chlorine	Cl	35·5	Potassium	K	39·1
Chromium	Cr	52·5	Silicon	Si	28
Cobalt	Co	59	Silver	Ag	108
Copper	Cu	63·5	Sodium	Na	23
Fluorine	F	19	Strontium	Sr	87·5
Gold	Au	196·6	Sulphur	S	32
Hydrogen	H	1	Tin	Sn	118
Iodine	I	127	Tungsten	W	184
Iron	Fe	56	Zinc	Zn	65

TABLE A.

EXERCISE I.—(See (10) for Examples for Practice.)

1. Analysis of Substances containing a Single Metal.—Determination of the Class to which the Metal present belongs (2).

TABLE A.

Dissolve (3) the substance by boiling with WATER, or HYDROCHLORIC ACID (dil.), or NITRIC ACID (dil.), or a mixture of HYDROCHLORIC ACID with a little NITRIC ACID. If the solution be not clear, filter it (4). The clear liquid is called the *original solution*.

To a part of the solution add HYDROCHLORIC ACID (dil.) (5.) [If no precipitate,* see next column]. Precipitate may contain— Lead, Silver, Mercury. See Table B.	To the same part, add HYDROSULPHURIC ACID† in excess (5). [If no precipitate, or only *white* sulphur (7), even on boiling the liquid (fig. 4), see next column]. Precipitate may contain— Lead, Bismuth, Copper, Mercury, Tin, Arsenic, Antimony. See Table C.	To a fresh part of the solution, add CHLORIDE OF AMMONIUM (8), AMMONIA‡ in excess (3), and SULPHIDE OF AMMONIUM (Hydrosulphate of Ammonia) [If no precipitate, see next column]. Precipitate may contain— If Black— Iron, Cobalt, Nickel. If not Black— Manganese, Zinc, Aluminium, Chromium, Phosphate of Lime. See Table D.	To a fresh part of the solution, add CHLORIDE OF AMMONIUM (8), AMMONIA in excess, and CARBONATE OF AMMONIA in small quantity. [If no precipitate, even on boiling, see next column]. Precipitate may contain— Calcium, Barium, Strontium. See Table E.	To the same part, add PHOSPHATE OF SODA, Na_2HPO_4. [If no precipitate, even on stirring (6), see next column]. Precipitate contains— *Magnesium* (9).	Metals not precipitated by the preceding tests:— *Potassium*, *Sodium*, *Ammonium*. See Table F.

* A precipitate is known to be formed if the liquid becomes partly or entirely opaque, in consequence of the separation of particles in a solid state.
† Be sure that the Hydrosulphuric Acid has a strong smell of the gas, since it becomes useless when long kept. If much nitric acid has been employed in dissolving the substance, it is advisable to evaporate down and to dilute with water, before adding hydrosulphuric acid.
‡ Even should Ammonia cause a precipitate, the hydrosulphate of ammonia must be added, as if no precipitate had been produced. Aluminium is liable to be missed here, on account of the transparent character of its precipitate; by warming the solution, the precipitate is rendered more visible.

EXPLANATIONS AND INSTRUCTIONS ON TABLE A.

2. For analytical purposes, the metals are classified in groups, according to the behavior of solutions containing them, on the addition of certain tests.

Analytical Classification of the Metals.

I.	II.	III.	IV.	V.
Metals, the solutions of which are precipitated by HYDROCHLORIC ACID, HCl, because their compounds with chlorine (chlorides) are insoluble, or nearly so, in water and in diluted acids.	Metals, the solutions of which are precipitated by HYDROSULPHURIC ACID, H_2S, in the presence of hydrochloric acid, because their compounds with sulphur (sulphides) are insoluble in water, and in cold diluted acids.	Metals, the solutions of which are precipitated by HYDROSULPHATE OF AMMONIA $(NH_4)_2S$, in the presence of ammonia, because the sulphides of the first five and the oxides of the last two are insoluble in water, and in ammonia or its salts.	Metals, the solutions of which are precipitated by CARBONATE OF AMMONIA, $(NH_4)_2CO_3$, because their carbonates are insoluble in water, and in ammonia or its salts.	Metals, the solutions of which are not precipitated by the foregoing tests.
Lead, Pb, *Silver*, Ag, *Mercury*, Hg (in the mercurous or proto-salts).	*Lead*, Pb, *Mercury*, Hg (in the mercuric or per-salts). *Bismuth*, Bi, *Copper*, Cu, *Tin*, Sn, *Antimony*, Sb, *Arsenic*, As.	*Iron*, Fe, *Nickel*, Ni, *Cobalt*, Co, *Manganese*, Mn, *Zinc*, Zn, *Aluminium*, Al, *Chromium*, Cr.	*Barium*, Ba, *Strontium*, Sr, *Calcium*, Ca.	*Magnesium*, Mg, *Potassium*, K, *Sodium*, Na, *Ammonium*, NH_4.

3. *To dissolve the substance for analysis,* place about five grains of the powdered substance (as much as can be taken easily on the end of the large blade of a pocket-knife) in a

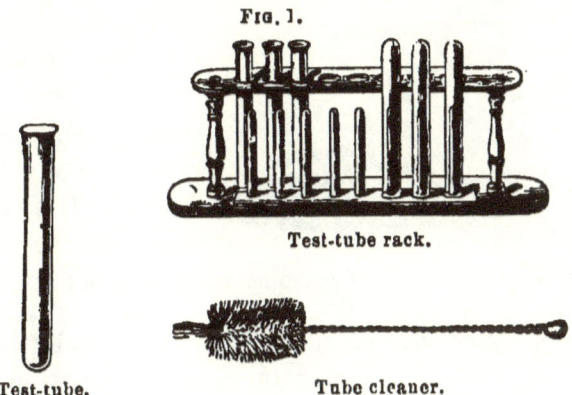

Fig. 1.

Test-tube rack.

Test-tube.　　　Tube cleaner.

test-tube (fig. 1), pour upon it about two drachms (two teaspoonfuls) of distilled water, shake them together, and if necessary, boil the water over the flame of a spirit-lamp (fig. 2) or a gas-burner; in the latter case, holding the tube a little above the flame, so as not to smoke it (figs. 3 and 4). The tube may be held in a band of folded paper when it is necessary to boil for some time.

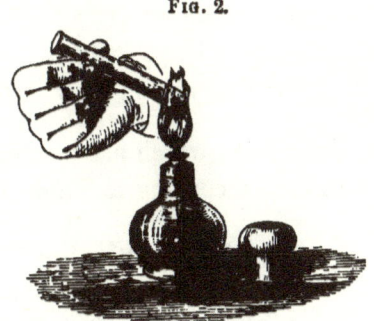

Fig. 2.

Fig. 3.　　　　　　Fig. 4.

If the substance does not appear to diminish in quantity, it may be set down as undissolved by the water.

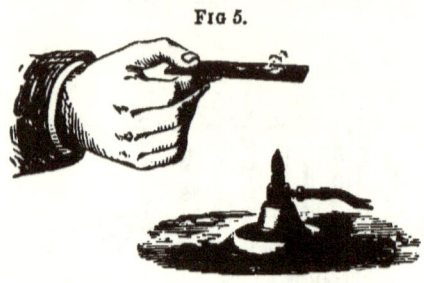

Fig 5.

Should there be any doubt, filter the solution (4), catch a drop of it upon a piece of thin window glass, and evaporate it at a gentle heat (fig. 5). If no considerable residue of solid matter is left, the substance may be considered insoluble.

Should water have failed to dissolve the substance, pour off the water so as to leave the powder, if possible, at the bottom of the tube; pour upon it about a drachm of diluted hydrochloric acid, and boil if necessary.

If the substance be insoluble in hydrochloric acid, boil a fresh portion, in another tube, with diluted nitric acid, and should this fail to dissolve it, add a few drops of nitric acid to the hydrochloric acid previously employed, and again boil.

Substances which are insoluble in water and acids must be examined according to Table I.

4. *To filter a solution.*—Take a circular piece of white filtering (blotting) paper, three or four inches in diameter, fold it neatly as in fig. 6, open it so as to form a cone, place it

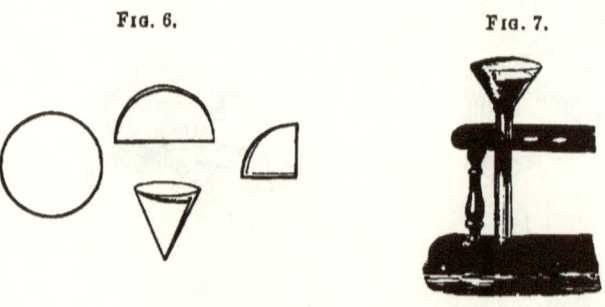

Fig. 6.   Fig. 7.

in a funnel, moisten it with distilled water, support the funnel in a test tube, as in fig. 7, and pour upon it the solution to be filtered. Should the funnel happen to fit air-tight into the test-tube, interpose a little piece of wood or paper to leave a passage for the air.

If the first filtration does not clear the liquid, it must be poured back through the same filter.

5. *Addition of tests to liquids.*—As a general rule, tests are gently poured, drop by drop, down the side of the test-tube (fig. 8), which is gently shaken, until either the expected effect is produced, or a reasonable proportion of the test has been added without any result. The stopper of the bottle

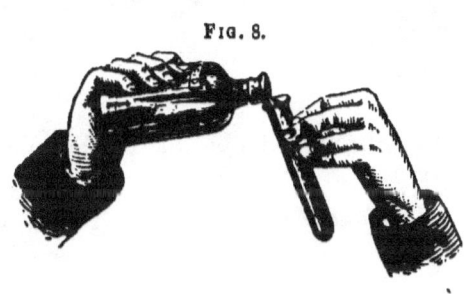

Fig. 8.

should not be laid upon the table, but should be held between the second and third fingers of the left hand, as in fig. 8, and the label of the bottle should always be upwards.

Excess.—In cases where the test is to be added *in excess*, the addition is continued until no further effect is produced by adding another portion, and until some conspicuous property of the test becomes evident in the liquid to which it is added.

Thus, Hydrosulphuric Acid is known to be in excess when a strong smell of it is perceived at the mouth of the test-tube after closing the tube with the thumb, and violently shaking it. Since the strongest hydrosulphuric acid only contains about three times its volume of sulphuretted hydrogen gas, it is necessary to add this test in large quantity, often amounting to three or four times the volume of the liquid to be tested. Again, Ammonia would be known to have been added in excess by its powerful odor, but if the liquid tested be strongly acid, it is not advisable to close the tube with the

strongly acid liquids is of a violent character. The solution may be mixed by pouring from one tube into another.

6. *Precipitation promoted by stirring*.—The formation of the precipitate of phosphate of ammonia and magnesia is much facilitated by stirring the liquid with a glass rod, or, still better, by rubbing the rod against the side of the tube (fig. 9), the latter being held in an inclined position, and the rod held *short* so that it cannot possibly be thrust through the bottom of the tube.

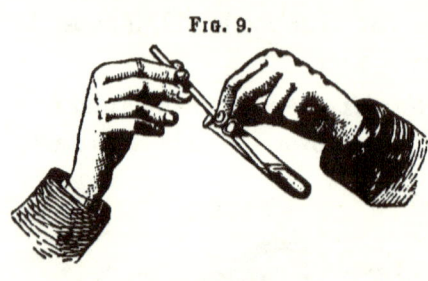
Fig. 9.

To make a stirring-rod.—The glass rod used for this purpose should be about six inches long, and rounded at both ends. To make it from one of the long glass canes sold at the glass-house, cut off six inches by making a deep scratch with a three-cornered file, and breaking the rod at this point by a sharp jerk (fig. 10). Fuse each end in the extreme point of the blowpipe flame (fig. 11) till well rounded. The glass must not be introduced into the inner part of the flame, or it will be permanently blackened, from the separation of lead in the metallic state.

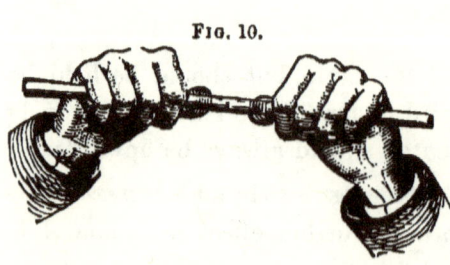
Fig. 10.

Fig. 11.

NOTES TO TABLE A.

7. The deposition of sulphur from the hydrosulphuric acid may be due to a variety of causes; for example, to the presence of nitrous acid derived from the nitric acid employed in dissolving the substance; to free chlorine; ferric chloride (perchloride of iron); sulphurous acid; chromic acid.

8. Chloride of ammonium is added to prevent the premature precipitation of magnesia, and should be added in considerable quantity. A solution containing a salt of magnesia yields, on addition of ammonia, a precipitate of hydrate of magnesia, but if chloride of ammonium be previously added, no precipitate is formed.

Should the chloride of ammonium itself produce a precipitate, before the ammonia is added, it probably consists of silica, and the substance under examination is likely to be silicate of potash or soda (77, 80).

9. Should the precipitate produced by phosphate of soda be *flocculent* instead of *granular* and *crystalline*, it is probably not caused by magnesium, but by aluminium or calcium which ought to have been detected in column 3 or 4.

In determining the particular form in which the magnesium is present in the substance under examination, assistance will be derived from the following statements.

Metallic Magnesium, Mg, is silver-white, burns easily in air, with a very brilliant light, and is dissolved, with effervescence, when boiled with chloride of ammonium.

COMMON COMPOUNDS OF MAGNESIUM.

Names.	Composition.
Calcined magnesia	Magnesium, oxygen.
Common magnesia, or Basic carbonate of magnesia	Magnesia, carbonic acid, water.
Epsom salts, or, Sulphate of magnesia	Magnesia, sulphuric acid, water.
Magnesite, or Carbonate of magnesia	Magnesia, carbonic acid.

Calcined Magnesia, MgO, is a white earthy powder, insoluble in water, and dissolved by hydrochloric acid, with little or no effervescence.

Basic Carbonate of Magnesia, $3MgCO_3,MgH_2O_2$, is similar, but effervesces rapidly with hydrochloric acid.

Magnesite, $MgCO_3$, is found in white earthy lumps which dissolve, with effervescence, in hydrochloric acid.

Sulphate of Magnesia, $MgSO_4,7Aq$, forms needle-like crystals, recognized by their bitter taste, easily dissolved by water, and copiously precipitated by nitrate or chloride of barium.

10. *Examples for Practice in Table A.*—The following substances will afford good practice in this Table, especially if they are presented to the student as puzzles, distinguished only by letters or numbers.

Sulphate of magnesia
Sulphate of iron
Acetate of lead
Arsenious acid
Carbonate of lime
Sulphate of zinc

Corrosive sublimate
Sulphate of copper
Carbonate of magnesia
Sulphide of antimony
Metallic tin.

TABLE B.   25

EXERCISE II.—(See (40) for Examples for Practice.)

11. Detection of a Metal belonging to the Hydrochloric Acid Group.

TABLE B.

EXAMINATION OF THE PRECIPITATE PRODUCED BY HYDROCHLORIC ACID.* (Add a few drops more HYDROCHLORIC ACID, to make sure that the precipitate will not redissolve.)

This precipitate may contain—

Chloride of Lead, Chloride of Silver, Mercurous Chloride.

Pour the liquid off the precipitate,† and shake the precipitate with AMMONIA.

DISSOLVED *Chloride of Silver*, Add dil. NITRIC ACID in excess, Precipitate *Chloride of Silver*, Presence of *Silver.* (12)	BLACKENED *Mercurous Chloride*, Presence of Mercury as mercurous compound. (13)	UNCHANGED *Chloride of Lead*, Presence of *Lead.*‡ (14)

* If the precipitate has a milky appearance, or a yellow color, and does not settle easily on standing, it is probably sulphur, in which case, the smell of hydrosulphuric or of sulphurous acid will be perceived at the mouth of the tube. (See *Hyposulphite of Soda* (80); also *Sulphide of Ammonium* (75), and *Sulphide of Potassium* (97).)

† If the precipitate refuses to subside, even after violent shaking, and does not appear to be sulphur, it may be poured upon a filter (4), the liquid allowed to pass through, and a few drops of ammonia poured over the precipitate.

‡ It is absolutely necessary to employ other tests for lead, as directed in (14), because hydrochloric acid occasions a precipitate in solutions of the silicates, stannates, and tungstates of the alkalies. Should lead not be discovered, another part of the original solution must be boiled with hydrochloric acid in excess, filtered, and further examined according to Table A. (See also *Silicate of Potash* (77), *Tungstate of Soda* (80), *Stannate of Soda* (37).)

NOTES TO TABLE B.

12. *Confirmatory Tests for Silver*, to be applied to the original solution.

Potash produces a brown precipitate of oxide of silver.

Bichromate of Potash produces a red precipitate of chromate of silver.

Chloride of Ammonium produces a white precipitate of chloride of silver.

If the original substance be *Metallic Silver*, Ag, it will be recognized, partly by its external characters, partly by its refusing to dissolve in hydrochloric acid, but dissolving easily in nitric acid. If greater certainty be desired, it may be examined by the blowpipe (264).

COMMON COMPOUNDS OF SILVER.

Names.	Composition.
Nitrate of silver, or Lunar caustic	Oxide of silver, nitric acid.
Chloride of silver	Silver, chlorine.

Nitrate of Silver, $AgNO_3$, is sold either in flat, tabular, transparent crystals, or in opaque cylindrical sticks made by fusing the crystals. It is dissolved very readily by cold water, and if filter-paper be moistened with the solution, and exposed to light, especially to sunlight, it assumes a black metallic appearance.

Chloride of Silver, AgCl, is insoluble in water and in acids, and will therefore not come under consideration at present (128).

13. *Confirmatory Tests for Mercury as a mercurous compound*, to be applied to the original solution.

Potash produces a black precipitate of mercurous oxide, Hg_2O.

Metallic Copper becomes silvery from the deposition of mercury.

Protochloride of Tin (Stannous chloride), added in excess, produces a gray precipitate of finely divided mercury.

Metallic Mercury (quicksilver) would be known at once.

Common Mercurous Compounds.

Names.	Composition.
Mercurous chloride, or Calomel	Mercury, chlorine.
Mercurous nitrate, or Protonitrate of mercury	Mercurous oxide, nitric acid, water.

Calomel, $HgCl$ or Hg_2Cl_2, is commonly sold as a white heavy powder, with a very slight shade of yellow. Before being ground to powder, it forms a translucent fibrous mass. It is not dissolved by water or dilute hydrochloric acid, and not, unless boiled, by dilute nitric acid; a mixture of the two acids dissolves it more readily when boiled. Calomel is easily known by its becoming black when shaken with potash or with lime-water, and dark gray when shaken with ammonia (250).

Protonitrate of Mercury, $HgNO_3, Aq.$, is sold in transparent prismatic crystals, which are decomposed when treated with water, a yellow basic nitrate of mercury being separated.

14. *Confirmatory Tests for Lead*, to be applied to the original solution.

Hydrochloric Acid produces a white precipitate of lead chloride, which dissolves on boiling, and is deposited in fine needles on cooling.

Dilute Sulphuric Acid produces, especially on stirring, a white precipitate of sulphate of lead, not dissolved by an excess of the acid, and soon depositing as a powder at the bottom of the tube.

Hydrosulphuric Acid produces a purplish black precipitate of sulphide of lead.

Bichromate of Potash produces a yellow precipitate of

chromate of lead. If the original substance be *Metallic Lead*, Pb, it will be recognized by its softness, and by its making a black streak on paper; it is insoluble in hydrochloric acid, but dissolves when boiled with dilute nitric acid; the solution gives a white precipitate with excess of ammonia, and a black precipitate, with a purplish shade, on addition of hydrosulphuric acid.

Common Compounds of Lead.

Names.	Composition.
Litharge, Massicot	Lead, oxygen.
Minium, or Red lead	Lead, oxygen.
Peroxide of lead,	Lead, oxygen.
Acetate of lead, or Sugar of lead	Oxide of lead, acetic acid, water.
Carbonate of lead, or White lead	Oxide of lead, carbonic acid, water.
Chromate of lead, or Chrome yellow	Oxide of lead, chromic acid.
Nitrate of lead	Oxide of lead, nitric acid.
Chloride of lead	Lead, chlorine.
Iodide of lead	Lead, iodine.
Oxychloride of lead	Lead, oxygen, chlorine.
Sulphate of lead	Oxide of lead, sulphuric acid.
Sulphide of lead, or Galena	Lead, sulphur.

Massicot, or Oxide of Lead, PbO, is a *yellow* powder, which is insoluble in water, and becomes white when boiled with hydrochloric acid, being converted into chloride of lead, which partly dissolves, and is deposited in needle-like crystals on cooling. Diluted nitric acid dissolves massicot when gently heated.

Litharge is the same oxide of lead, which has been melted, and is sold in *pinkish-brown* or buff scales or powder. The action of water and acids upon it is similar to that upon massicot.

LEAD COMPOUNDS. 29

Minium, Pb_3O_4, is a *bright red* powder, which is not affected by water, but evolves the smell of chlorine when boiled with hydrochloric acid, and is slowly converted into chloride of lead. Dilute nitric acid only partly dissolves it, leaving a brown powder (peroxide of lead).

Peroxide of Lead, or *Binoxide of Lead*, PbO_2, is a *dark brown* powder, which is insoluble in water and in nitric acid, but dissolves slowly in boiling hydrochloric acid, giving off the smell of chlorine, and forming chloride of lead, which crystallizes in needles from the solution as it cools.

Acetate of Lead, $Pb2C_2H_3O_2 3Aq.$, forms white *needle-like crystals*, which have a faint odor and an intensely sweet taste. When treated with water it dissolves, but yields a milky solution, especially with common water, from the presence of a little carbonate of lead, formed from the carbonic acid contained in the water; nitric acid clears it up at once. When acetate of lead is heated on the point of a knife, or on a slip of glass, it melts, and gives off an inflammable vapor with a very peculiar smell, leaving a gray residue composed of carbon and minute globules of metallic lead, easily changing to yellow oxide of lead.

Perchloride of iron (ferric chloride), added to a solution of acetate of lead, produces a white precipitate of chloride of lead, and a red solution of ferric acetate, which is clearly seen after the precipitate has subsided.

White Lead, or Basic Carbonate of Lead, $2PbCO_3, PbH_2O_2$, is a heavy earthy powder, soon becoming gray when exposed to air, from the action of sulphuretted hydrogen. It is insoluble in water, and effervesces with hydrochloric acid, dissolving when heated, as chloride of lead, which crystallizes in needles on cooling. Dilute nitric acid easily dissolves carbonate of lead, with effervescence caused by the escape of carbonic acid gas. When heated on a knife, or slip of glass, it becomes yellow.

Chromate of Lead, $PbCrO_4$, is commonly sold as a *bright yellow* or *orange red* powder or cake—(the *orange chrome* is the basic chromate or dichromate of lead, $PbCrO_4,PbO$); but fused chromate of lead has a brown color. It is insoluble in water, but dissolves slowly when boiled with strong hydrochloric acid, evolving chlorine and yielding a green solution. If it be heated with hydrochloric acid and a little alcohol, a bright green solution of chlorides of chromium and lead is produced, which deposits crystals of chloride of lead on cooling. Nitric acid scarcely affects chromate of lead. If the yellow chromate of lead be heated on a knife, or slip of glass, its color changes to a red-brown.

Nitrate of Lead, $Pb2NO_3$, forms *hard white crystals*, having a sweet taste. It dissolves in water, but not easily unless heated. When heated on a knife, or slip of glass, it crackles or *decrepitates* violently. If previously powdered, to prevent its flying off, it evolves suffocating brown fumes (nitric peroxide), and leaves a yellow or red residue.

Chloride of Lead, $PbCl_2$, is a *white* powder, sometimes crystalline, which dissolves slowly when boiled with water, the solution easily depositing crystals on cooling.

Iodide of Lead, PbI_2, is a bright *yellow powder*, which dissolves sparingly in boiling water, but more readily on adding a little hydrochloric acid, yielding a colorless solution, which deposits brilliant golden scales on cooling. When boiled with nitric acid, iodide of lead gives off the purple vapors of iodine.

Oxychloride of Lead is either *white* (Pb_2OCl_2) or *bright yellow* ($Pb_8O_7Cl_2$), according to the mode of preparing it. It melts easily when heated, the white oxychloride becoming yellow. It is insoluble in water, but dissolves sparingly when boiled with hydrochloric acid, the solution depositing crystals of chloride of lead on cooling.

Sulphate of Lead, $PbSO_4$, is a *white* powder, which does not change when heated, is insoluble in water, but dissolves slowly in boiling hydrochloric acid, the solution depositing crystals of chloride of lead as it cools.

Sulphide of Lead or *Galena*, PbS, is a *dark, gray, heavy, metallic-looking* substance, the masses of which may be easily split or *cleaved* with a knife-blade into rectangular fragments. Water has no effect upon it, and diluted hydrochloric acid very little, but nitric acid gradually dissolves it, brown fumes being evolved, resulting from the decomposition of the nitric acid. Strong hydrochloric acid also dissolves it on heating, producing the offensive odor of hydrosulphuric acid.

Phosphate of Lead, $Pb_3 2PO_4$, is described at (112).

15. Detection of a Metal belonging to the Hydrosulphuric Acid Group.

TABLE C.

EXAMINATION OF THE PRECIPITATE PRODUCED BY HYDROSULPHURIC ACID.
This precipitate may contain—

If Black or Brown.

Sulphide of Lead (19), *Mercuric Sulphide* (20), *Sulphide of Copper* (21), *Sulphide of Bismuth*, *Sulphide of Tin*.

Test separate portions of the original* solution with

1. SULPHURIC ACID. White Precipitate (24), *Sulphate of Lead*. Presence of *Lead* (14).

2. Twice its bulk of WATER, Milky Precipitate (*Bismuth Oxychloride*). Presence of *Bismuth* (25, 29). If clear, add MERCURIC CHLORIDE. White or Gray Precipitate. Presence of *Tin* as a stannous compound (22, 23).

Blue color. Presence of *Copper* (26, 27).

Add much WATER, Milky Precipitate. *Oxychloride of Bismuth*. Presence of *Bismuth* (28, 29).

3. AMMONIA in excess (5, 25).

White Precipitate indicates *Bismuth* or *Mercury*. Dissolve in *very little* dil. HYDROCHLORIC ACID, and div'de into three parts.

1. Add much WATER. Milky Precipitate. *Oxychloride of Bismuth*. Presence of *Bismuth* (28, 29).

2. Add IODIDE OF POTASSIUM. Red Precipitate. Presence of *Mercury* (30, 31). Add *a drop of* ACETATE OF LEAD. Brown Precipitate. Presence of *Bismuth* (28, 29).

3. Introduce a few strips of clean COPPER, and boil. Silvery coating. Presence of *Mercury*, as a mercuric compound (30, 31).

4. POTASH in excess (5). Yellow Precipitate. *Mercuric Oxide*. Presence of *Mercury*, as a mercuric compound (30, 31). If no precipitate, test the original solution by boiling with hydrochloric acid and copper (30, 31).

If Yellow or Orange-red.†

Sulphide of Arsenic, *Sulphide of Antimony*, *Bisulphide of Tin*. Heat a little of the precipitate with CARBONATE OF AMMONIA (32).

Dissolved. *Sulphide of Arsenic*. Probable presence of *Arsenic* (33, 34, 35).

Undissolved.

Yellow. *Bisulphide of Tin*. Probable presence of *Tin*, as a stannic compound (36, 37).

Orange red. *Sulphide of Antimony*. Probable presence of *Antimony* (38, 39).

---

* By the *original* solution is meant that to which no test has yet been added.

† In solutions of lead containing much hydrochloric acid, a bright red precipitate of chlorosulphide of lead is often produced by hydrosulphuric acid. This becomes black by contact with excess of hydrosulphuric acid, especially on heating.

EXPLANATIONS AND INSTRUCTIONS ON TABLE C.

16. *To wash a precipitate.* —The precipitate having been collected upon a filter (4) may be washed by filling the filter twice or thrice with distilled water, and allowing it to run through. A washing-bottle (fig. 12) will be of great assistance in washing precipitates, the stream of water being directed so as to wash the precipitate from the sides towards the apex of the filter.

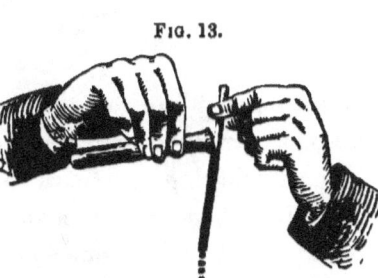

Fig. 12. 
Washing-bottle.

Fig. 13.

Fig. 14.

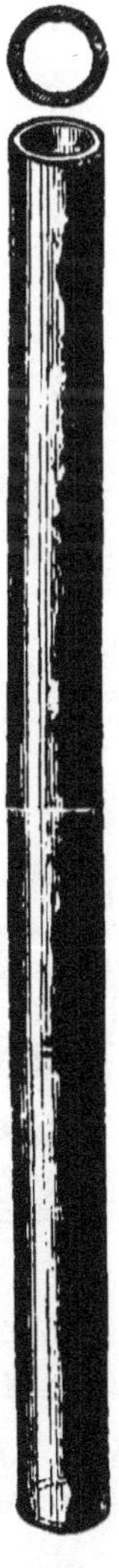

When a precipitate is very heavy and subsides readily, it may be *washed by decantation*—that is, by shaking it up with successive quantities of distilled water, which are poured off when the precipitate has settled down. A wet glass rod, held against the lip of the test-tube (fig. 13), greatly assists in decanting the liquid without disturbing the precipitate.

17. *Heating solid substances in tubes.* —Ordinary test-tubes should not be used for this purpose, but much smaller tubes, which are made, with the help of the blowpipe, from a piece of (German) glass tube free from lead, of the size represented in fig. 14. The middle of this piece of tube is softened in the blowpipe flame (fig. 15), and quickly drawn out to form two tubes connected by a mere thread of glass, which is then detached from each tube, as shown at *a*, leaving the finished tube of the shape shown at *b*.

FIG. 15.

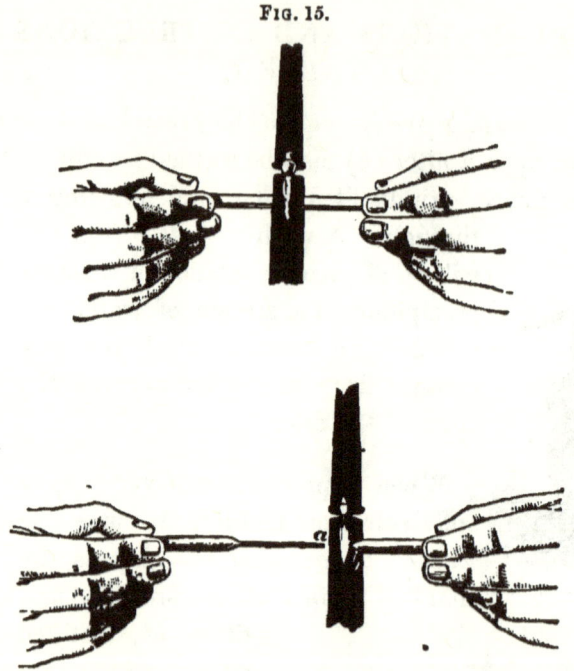

The substance to be heated is introduced into this tube, in very small quantity, and the sides of the tube are cleansed from adhering particles by rubbing them with a match. A strap of folded paper (fig. 16) is put round the tube, which is then held in the lower part of the flame of a gas-jet (to avoid smoking it), or in a spirit-flame. If any moisture should condense on the sides of the tube, incline its mouth downwards, so that the drops may not run back upon the heated glass and crack it. If it be necessary to apply a more intense heat, the blowpipe flame may be directed on to the bottom of the tube.

FIG. 16.

18. *To ascertain whether a solution is alkaline or acid,*

dip a piece of red or blue litmus-paper into it, or touch the paper with a glass rod dipped in the liquid. An alkaline solution changes red litmus to blue, whilst an acid solution reddens the blue litmus-paper. In the absence of test-paper the taste may of course be relied on for a rough indication.

NOTES TO TABLE C.

19. Lead is liable to be found in the precipitate produced by hydrosulphuric acid, because its chloride is dissolved to a considerable extent by water, so that a weak solution of a salt of lead is not precipitated by hydrochloric acid.

20. The mercuric salts generally give, on addition of a little hydrosulphuric acid, a yellow or brownish precipitate, which passes through various shades on adding more hydrosulphuric acid, finally becoming black.

21. Solutions of copper have a blue or green color. The blue solutions are turned green by hydrochloric acid.

22. The perchloride of mercury (mercuric chloride) will give, with stannous salts or protosalts of tin, a white precipitate of mercurous chloride (calomel) or a gray precipitate of finely-divided metallic mercury, accordingly as one-half or the whole of the chlorine is abstracted from the mercuric chloride by the protosalt of tin.

23. *Metallic Tin*, Sn, would be easily known by its becoming converted into a white powder (binoxide of tin) when boiled with nitric acid. It dissolves slowly when boiled with diluted hydrochloric acid, and rapidly in the strong acid yielding a solution of protochloride of tin (or stannous chloride), which is at once recognized by the test with perchloride of mercury mentioned above (22).

For the common protosalts of tin, see the next page.

COMMON STANNOUS COMPOUNDS, OR PROTOSALTS OF TIN.

Names.	Composition.
Protochloride of tin or Salts of tin, or tin crystals	Tin, chlorine, water.
Protosulphide of tin, or Tin pyrites	Tin, sulphur.

Protochloride of Tin, $SnCl_2, 2Aq.$, usually forms needle-like crystals with a tinge of yellow. Water decomposes it, forming a milky-white oxychloride of tin. Hydrochloric acid dissolves it entirely.

Tin Pyrites, SnS, is a dark gray mineral which is slowly dissolved when boiled with hydrochloric acid, evolving the smell of hydrosulphuric acid. Boiling nitric acid converts it into the white binoxide of tin.

24. The precipitate of sulphate of lead sometimes forms only after the lapse of some minutes in highly acid solutions.

25. If the quantity of the original solution is very limited, ammonia may be added in excess to that portion which has been already tested with sulphuric acid.

26. The presence of copper may be confirmed by introducing a piece of clean iron or steel into this solution mixed with a slight excess of hydrochloric acid; a red coating of metallic copper will be deposited upon the metal.

Ferrocyanide of potassium may be added to the original solution; if copper is present, a red-brown precipitate of ferrocyanide of copper will be produced.

27. *Metallic Copper*, Cu, is identified by its color, its insolubility in diluted hydrochloric acid, and by its dissolving rapidly in nitric acid, to a green or blue solution of nitrate of copper.

Common Compounds of Copper.

Names.	Composition.
Sulphate of copper, or Blue-stone	Oxide of copper, sulphuric acid, water.
Arsenite of copper, or Scheele's green	Oxide of copper, arsenious acid, water.
Basic acetate of copper, or Verdigris	Oxide of copper, acetic acid, water.
Basic carbonate of copper, or Malachite	Oxide of copper, carbonic acid, water.
Oxychloride of copper, or Brunswick green	Oxide of copper, chloride of copper, water.
Oxide of copper, or Cupric oxide	Copper, oxygen.
Suboxide of copper, or Cuprous oxide	Copper, oxygen.
Sulphide of copper	Copper, sulphur.

Sulphate of Copper, $CuSO_4, H_2O, 4Aq.$, forms *blue diamond-shaped crystals*, often grouped into irregular masses. It dissolves easily in water, yielding a blue solution, which is copiously precipitated by nitrate or chloride of barium. Dried sulphate of copper is nearly white, and becomes blue when moistened.

Arsenite of Copper, $CuHAsO_3$, is a *bright green powder*, insoluble in water, but soluble in hydrochloric acid to a green liquid. It may be tested for arsenious acid according to (33).

Verdigris, $Cu2C_2H_3O_2, CuO, 6H_2O$, has a *bluish green* color. It is partly dissolved by water, and entirely by hydrochloric acid, to a green solution. When heated on a knife or a slip of glass, it is blackened and emits an odor like that of vinegar.

The *Basic Carbonates of Copper* are *blue* ($2CuCO_3, Cu H_2O_2$) or *green* ($CuCO_3, CuH_2O_2$), insoluble in water, but soluble, with effervescence, in hydrochloric acid.

Oxychloride of Copper, $Cu_4Cl_2O_3,4Aq.$, is *green*. It is insoluble in water, but dissolves in acids. Its solution in nitric acid may be tested for chlorine with nitrate of silver (Table H).

Cupric Oxide, CuO, is a *black* powder, insoluble in water, but soluble in hydrochloric acid, on boiling, forming a green solution, which sometimes becomes turbid when mixed with water, from the separation of a little subchloride of copper.

Cuprous Oxide, Cu_2O, is *red* or *red-brown;* it is not dissolved by water, but hydrochloric acid dissolves it, on boiling, to a brown solution, which gives a thick white precipitate of cuprous chloride (subchloride of copper) when mixed with water. Boiling nitric acid dissolves cuprous oxide, forming a blue solution which is not precipitated by water.

Sulphide of Copper, Cu_2S, as found in nature (cuprous sulphide or copper-glance), is a *black* substance with a somewhat metallic lustre; insoluble in water and in hydrochloric acid, but dissolved by boiling in nitric acid to a blue solution, spongy flakes of dark colored sulphur usually separating.

The artificial sulphide of copper, CuS, is usually a black powder with a shade of green, which behaves in the same way.

28. If the precipitate of oxide of bismuth, Bi_2O_3, produced by ammonia be small, it should be collected upon a filter, washed with a little water (16), and very carefully dissolved by dropping two or three drops of diluted hydrochloric acid upon it. When water is added to the slightly acid solution of chloride of bismuth so produced, a milky precipitate of oxychloride of bismuth is formed, but the presence of an excess of acid prevents its formation.

To confirm the presence of bismuth, add to the original solution (or to the solution of the oxide precipitated by ammonia, in hydrochloric acid, even after dilution) solution of iodide of potassium; this will produce a red or yellow color if bismuth be present (or even, in a strong solution, a brown precipitate of iodide of bismuth), and on adding a drop of acetate or nitrate of lead, the iodide of lead which precipitates will have a brown or red instead of its usual yellow color, from the presence of iodide of bismuth. If this precipitate be dissolved by heating the liquid and adding a few drops of dil. hydrochloric or acetic acid, it will crystallize out in very beautiful brown or red scales as the solution cools.

An excellent confirmatory test for bismuth consists in adding to the original solution some protochloride of tin (stannous chloride, prepared by boiling a fragment of tin with strong hydrochloric acid) and an excess of potash, when a black precipitate of bismuthous oxide, BiO, is obtained. Should the stannous chloride produce a gray or white precipitate, becoming dark gray on adding potash, it is probably due to the presence of mercury.

29. *Metallic Bismuth*, Bi, is a brittle metal with a faint pink reflection. It does not dissolve in hydrochloric acid, but dissolves in diluted nitric acid, on boiling.

Common Compounds of Bismuth.

Names.	Composition.
Basic nitrate of bismuth, or Flake white	Oxide of bismuth, nitric acid, water.
Oxychloride of bismuth, or Pearl white	Oxide of bismuth, chloride of bismuth, water.

Flake white,* $Bi_2N_2O_8.H_2O$, is insoluble in water, but dissolves in hydrochloric acid. When heated in a dry glass tube (17), it evolves moisture, which condenses in drops on the cool part of the tube, and brown vapors of nitric peroxide.

Pearl white, BiOCl, is also insoluble in water and soluble

---

* Flake white sometimes consists of basic carbonate of lead (white lead).

in hydrochloric acid. It may be dissolved in nitric acid, and tested for chlorine with nitrate of silver (Table II).

30. In order to be quite sure of the presence of mercury in a solution, it must be boiled with metallic copper. If nitric acid be present, which would dissolve the copper, it must be neutralized by adding a slight excess of ammonia (5); enough dilute hydrochloric acid must then be added to destroy the odor of ammonia, even after shaking, and two or three slips of bright copper introduced. On boiling, the copper will acquire a bright silvery coating, and if it be rinsed with water, dried in filter-paper and heated in a dry tube (17), a gray crust of minute globules of mercury will be formed upon the side of the tube, uniting into larger globules when rubbed with a glass rod or a wooden match. Bismuth also deposits upon the copper, but forms a *dull* gray coating.

When mercury is present in the form of cyanide of mercury, it would generally escape detection until the boiling with hydrochloric acid and copper, at the end of Table C.

31. Common Mercuric* Compounds.

Names.	Composition.
Mercuric chloride, or perchloride, or Bichloride of mercury, or Corrosive sublimate	Mercury, chlorine.
Mercuric sulphide, or Vermilion, or cinnabar	Mercury, sulphur.
Mercuric oxide, or Nitric oxide of mercury	Mercury, oxygen.
White precipitate	Mercury, chlorine, nitrogen, hydrogen.
Mercuric iodide	Mercury, iodine.
Mercuric cyanide	Mercury, cyanogen (carbon, and nitrogen).

* Nitric acid converts mercurous compounds into mercuric compounds; hence, if this acid should have been used to dissolve the original substance, the latter may possibly have been a mercurous

Corrosive Sublimate, $HgCl_2$, is sold either in *shining white semi-transparent masses* or as a white crystalline powder. It dissolves readily when boiled with water, and crystallizes from a strong solution in white needles. When heated in a small tube (17), it melts very easily to a perfectly clear liquid, which crystallizes in fine needles on cooling. By continuing the heat, it is boiled away in vapor which has a fearfully suffocating effect upon the nose.

Vermilion, HgS, is known by its *bright red* color. It is insoluble in water, and in hydrochloric or nitric acid separately, but it dissolves in a mixture of the two acids, with separation of spongy flakes of sulphur.

Cinnabar, HgS, the chief ore of mercury, is generally met with in *dark brown* very heavy hard masses, which become red when scraped with a knife. In its relation to solvents it resembles vermilion.

Nitric Oxide of Mercury, HgO, is a *bright red* shining powder, insoluble in water, but soluble in hydrochloric acid.

White Precipitate, $HgClNH_2$, is a heavy white earthy-looking substance, insoluble in water, but soluble in hydrochloric acid. When boiled for some time with water, it becomes yellow. On boiling it with potash, it evolves the odor of ammonia.

Mercuric Iodide, or Biniodide of Mercury, HgI_2, is a *scarlet* powder, insoluble in water, but dissolved by boiling hydrochloric acid. When heated on a slip of glass, it becomes bright yellow, and passes off in yellow fumes. The yellow powder becomes red when rubbed with a glass rod.

Mercuric Cyanide, or Cyanide or Bicyanide of Mercury, $Hg2CN$, forms white prismatic crystals which dissolve in boiling water. When heated in a dry tube (17) they crackle, melt, and evolve cyanogen, which may be recognized by its odor and by its burning with a pink flame. A brown residue is left at the bottom of the tube, and the cooler part of the tube is covered with a gray deposit of mer-

32. The carbonate of ammonia is to be added in excess, so that the smell of ammonia is quite perceptible on warming the liquid; otherwise the sulphide of arsenic would not be dissolved and might be mistaken for that of tin.

It is better to collect a little of the precipitate upon a filter, to wash it once or twice with water (16), and to pour some warm carbonate of ammonia over it. If there be any doubt whether the precipitate has been dissolved, test a little of the solution in carbonate of ammonia with excess of hydrochloric acid, when sulphide of arsenic will be separated in yellow flakes.

33. Should any confirmatory test be required for arsenic, either Reinsch's test or Marsh's test may be employed.

Reinsch's test for arsenic.—Boil a little of the original substance with excess of hydrochloric acid and a few strips of bright copper for a minute or two. The copper displaces the arsenic from the solution, and a dark gray compound of arsenic with copper is formed upon the surface of the strips. Rinse these with a little water, dry them on filter-paper, and heat them gently in a small tube closed at one end (17); minute shining crystals of arsenious acid will be deposited on the cool part of the tube, having been produced by the combination of the arsenic with oxygen from the air. If the crystalline deposit is examined with the microscope (for which purpose the binocular is to be preferred), it will be seen to consist of octahedral crystals (fig. 17).

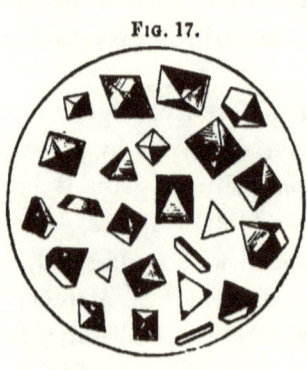

Fig. 17.

34. *Marsh's test for arsenic.*—Dissolve the substance, if possible, in water or hydrochloric acid. If it be insoluble in these, dissolve it in nitric acid, evaporate the solution to dryness in a small dish, and redissolve the residue in water with

MARSH'S TEST FOR ARSENIC. 43

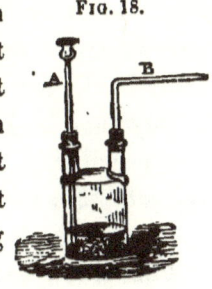

Fig. 18.

Arrange an apparatus as represented in fig. 18; the funnel-tube A, and the bent tube (229) B, being passed through air-tight perforated corks (228). B has been drawn out to a moderately fine jet by softening it in the blowpipe flame (fig. 15), drawing it out to a narrow neck (fig. 19), and cutting this across with a file, at *a*.

Fig. 19.

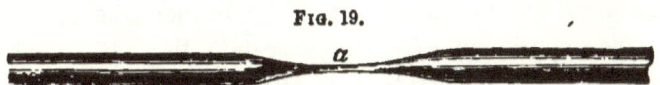

Introduce into the bottle enough granulated zinc* to cover the bottom, fill the bottle about one-third with water, and pour in dilute sulphuric acid, through the funnel-tube (the lower end of which must dip beneath the water) until moderate effervescence, from the evolution of hydrogen gas, takes place. Keep the tube as far away from a flame, because the bottle is now filled with an explosive mixture of hydrogen and air. Incline the gas-bottle and hold a small test-tube over the tube *b*, as represented in fig. 20, for about a minute, slowly withdraw it, keeping its mouth downwards, and apply it to a flame; if it explodes, the hydrogen in the bottle is still

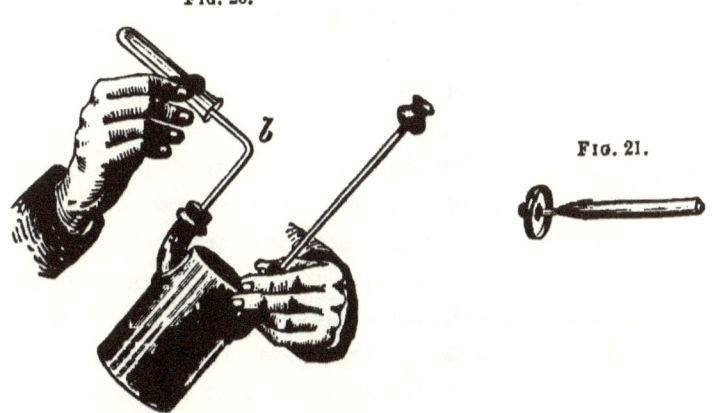

Fig. 20.

Fig. 21.

* To granulate zinc, melt it in an iron ladle, stand upon the

mixed with air, and it would explode if lighted. Repeat the experiment until the hydrogen can be seen to burn steadily away in the test-tube. The hydrogen issuing from the tube B may then be kindled with safety. Hold the lid of a porcelain crucible in the flame (fig. 21), pressing it close up to the jet. No spot (of arsenic or antimony) will be deposited upon the porcelain if the hydrogen be pure.

Pour a few drops of the solution to be tested for arsenic, prepare as directed above, into the funnel-tube A, and repeat the experiment with the porcelain lid. If arsenic be present, arsenietted hydrogen will be formed, and will deposit a dark brown stain of arsenic on the porcelain surface.

Make a stain upon another lid in the same way.

Fig. 22.

Heat the tube B with a spirit-lamp (fig. 22) for a minute or two, when the arsenietted hydrogen will be decomposed, and a nearly black shining crust of arsenic will be deposited on the cooler part of the tube.

To prove that these results are due to arsenic, and not to antimony, which might imitate them, touch one of the stains with a glass rod dipped in solution of chloride of lime, which will dissolve a stain of arsenic, but not that of antimony.

Test the other stain in the same way with yellow hydrosulphate of ammonia, which will not dissolve a stain of arsenic, but dissolves that of antimony. If the deposit formed in B were due to antimony, it would be produced close to the hot part of the tube, whilst the arsenical crust is deposited upon the cooler part of the tube. To be quite sure of its character, make a deep file-mark at each end of it, and break off the end of the glass. Wrap the piece containing the deposit in a piece of stout paper, break it into fragments (not into powder), and heat one or two of these in a small tube (17), when the arsenic will be oxidized, and a shining ring of crystals of arsenious acid will be deposited on the cooler part of the tube.

35. *Metallic Arsenic*, As, is a *dark gray* brittle substance, with metallic lustre. It is insoluble in water and in hydrochloric acid, but dissolves in boiling nitric acid. When heated in air, it emits a smell of garlic. Heated in a dry tube (17) it is converted into vapor, which condenses higher up the tube, partly as a black shining crust, partly as a white crystalline powder of arsenious acid produced by the oxygen of the air in the tube.

Common Compounds of Arsenic.

Names.	Composition.
Arsenious acid, or White arsenic	Arsenic, oxygen.
Arsenite of copper, or Scheele's green	Arsenious acid, oxide of copper.
Arsenic acid	Arsenic, oxygen.
Arseniate of soda	Arsenic acid, soda, water.
Tersulphide of arsenic, or Yellow orpiment	Arsenic, sulphur.
Bisulphide of arsenic, or Realgar or red orpiment	Arsenic, sulphur.
Iodide of Arsenic	Arsenic, iodine.

Arsenious Acid, As_2O_3, is sold either in *opaque porcelain-like masses*, or as a *white powder*. It seems insoluble in water, unless long boiled with it, but dissolves in boiling hydrochloric acid, and more readily in boiling nitric acid, evolving from the latter brown fumes of nitrous acid.

When a little arsenious acid is heated in a dry tube (17), it is entirely converted into vapor, and is deposited on the cooler part of the tube as a white crystalline powder (246, 251).

Arsenite of Copper has been described at page 37. When

heated in a dry tube (17), it evolves vapor of arsenious acid which condenses to a crystalline powder.

Arsenic Acid, As_2O_5, is commonly sold in combination with water, as hydrated arsenic acid, in *white irregular lumps*, which soon become damp by absorbing water from the air. It dissolves in water, and if the solution be mixed with hydrochloric acid, and a considerable volume of hydrosulphuric acid added, it is not precipitated in the cold, but when boiled it yields a white precipitate of sulphur, followed by a yellow precipitate of tersulphide of arsenic.

Arseniate of soda, $Na_2HAsO_4.7Aq.$, forms prismatic crystals which dissolve easily in water, yielding a solution which is alkaline to test-papers (18), and behaves like arsenic acid with hydrochloric and hydrosulphuric acids.

The solution of arseniate of soda gives a red-brown precipitate with nitrate of silver, and the solution of arsenic acid will give the same precipitate if ammonia be *very cautiously* added after the nitrate of silver.

Tersulphide of Arsenic, As_2S_3, is *bright yellow*, insoluble in water and in hydrochloric acid, but dissolved by boiling nitric acid, with separation of sulphur. It dissolves in warm ammonia, or potash, yielding a colorless solution, from which it is reprecipitated in yellow flakes by an excess of hydrochloric acid.

Realgar, As_2S_2, is sold in *orange-red* masses or powder. It behaves like the preceding with water and acids, but is not entirely dissolved by potash, leaving a dark brown residue.

Iodide of Arsenic, AsI_3, forms *brick-red* flakes which are dissolved to a considerable extent by boiling water, and evolve violet vapors when boiled with nitric acid.

36. If too little carbonate of ammonia be added, the sulphide of arsenic may be mistaken for bisulphide of tin. On adding ammonia, the sulphide of arsenic dissolves very readily, but the bisulphide of tin dissolves with difficulty.

To confirm the presence of tin, place in a little of the original solution (which should contain free hydrochloric but not nitric acid) a piece of zinc; after a short time metallic tin will be deposited as a spongy mass; rinse this with water, boil it with a little hydrochloric acid to dissolve it in the form of chloride of tin (stannous chloride), and test it with perchloride of mercury (22).

If the original solution contains nitric acid, add ammonia in slight excess to neutralize it, and acidify the solution again with hydrochloric acid before testing with zinc.

37. The mode of identifying *metallic tin* has been described at (23).

COMMON STANNIC COMPOUNDS (OR PERSALTS OF TIN.)*

Names.	Composition.
Binoxide of tin, or Stannic acid	Tin, oxygen.
Stannate of soda	Stannic acid, soda, water.
Bisulphide of tin *Aurum musivum*	Tin, sulphur.
Stannic chloride, or Nitromuriate of tin	Tin, chlorine (water).
Pink salt	Stannic chloride. Hydrochlorate of ammonia.

Binoxide of Tin, SnO_2, is insoluble in water and acids and will, therefore, not be considered at present.

Stannate of Soda, $Na_2SnO_3.4Aq.$, is usually sold in *opaque irregular crystals*, which are soluble in cold water, though generally leaving a white residue. The solution is alkaline (18). Hydrochloric acid added drop by drop causes a white precipitate of stannic acid, which is redissolved by an excess of the acid.

Bisulphide of Tin, SnS_2, or bronze powder, or Mosaic

---

* Nitric acid converts stannous compounds into stannic compounds; hence, if this acid has been used to dissolve the original substance, this may have been a stannous compound (23).

gold, is a *golden yellow*, scaly substance, insoluble in water, in hydrochloric acid, and in nitric acid, but dissolved by boiling with hydochloric acid and adding a little nitric acid. Much nitric acid causes the separation of white binoxide of tin.

Stannic Chloride, $SnCl_4$, is commonly met with in the state of solution, which is highly acid to test-papers (18). It may be tested for chlorine with nitrate of silver (Table II).

Pink Salt, $SnCl_4.2NH_4Cl$, may be tested for chlorine in a similar manner. When heated with potash it evolves ammonia.

38. The presence of antimony may be confirmed by acidifying the original solution with hydrochloric acid, and introducing a piece of zinc, when a sooty black powder of metallic antimony will be deposited.

Or a few drops of the solution (free from nitric acid) acidified with hydrochloric acid, may be placed upon a surface of platinum (foil) and a piece of zinc made to touch the platinum; a dark stain of antimony will be formed upon the latter metal. If this stain be rinsed with water, and wetted with yellow hydrosulphate of ammonia, it will be dissolved on warming, and the solution, if evaporated on the platinum, will deposit the orange-colored sulphide of antimony.

Or the solution, containing hydrochloric acid only, and not nitric acid, may be tested by Marsh's test (34).

Or a little of the original substance may be boiled with excess of hydrochloric acid, and a few strips of copper, when the latter will displace the antimony and become covered with a purple antimonial film.

39. *Metallic Antimony*, Sb, is known by its great brittleness and brilliant lustre. It is not attacked by water, and to a slight extent only by boiling hydrochloric acid. When boiled with diluted nitric acid, it is converted, though slowly, into a white powder (antimonic acid) which is slightly soluble in the nitric acid. A mixture of hydrochloric acid with a little nitric acid dissolves the metal, and if a large excess of acid be avoided, the addition of a large volume of water to the solution causes a thick white precipitate of oxychloride of antimony.

Common Compounds of Antimony.

Names.	Composition.
Tersulphide of antimony, or Crude antimony ore	Antimony, sulphur.
Teroxide of antimony Flowers of antimony	Antimony, oxygen.
Tartar-emetic	Teroxide of antimony. Potash, tartaric acid, water.
Terchloride of antimony	Antimony, chlorine.
Antimoniate of potash	Antimonic acid, potash.

Tersulphide of Antimony, Sb_2S_3, as found in nature, is a *dark gray* crystalline substance with decidedly metallic lustre. It is unaffected by water, but boiling hydrochloric acid dissolves it slowly, evolving the odor of hydrosulphuric acid. If the solution thus obtained be filtered and mixed with water, it generally gives an orange-red precipitate. A mixture of hydrochloric acid with a little nitric acid dissolves it readily, flakes of sulphur being separated.

The artificial tersulphide of antimony, or antimony-vermilion, is an *orange red* powder, which behaves with acids like the native tersulphide.

Teroxide of Antimony, Sb_2O_3, is a *grayish-white* powder, insoluble in water, but dissolved by hydrochloric acid. When heated on a knife or a slip of glass, it becomes yellow, but turns white again on cooling.

Tartar-emetic $2(K.SbO.C_4H_4O_6).Aq.$, forms hard white crystals, or a white powder, readily dissolved by hot water. The solution gives a white precipitate of teroxide of antimony on adding a drop of diluted hydrochloric acid, but an excess of acid readily dissolves it. When tartar-emetic is heated on a knife or a slip of glass it is carbonized, and evolves the peculiar odor of burnt sugar, which characterizes the products of decomposition of tartaric acid.

Terchloride of Antimony, $SbCl_3$, in the pure state forms a soft gray fusible solid, but it is commonly met with in the state of solution, usually of a yellow color, due to the presence of iron. When largely diluted with water, it gives a white precipitate of oxychloride of antimony.

Antimoniate of Potash, $KSbO_3$, is usually sold as a white powder, which is partly dissolved by boiling water, yielding an alkaline solution (18). If this solution be filtered, and a drop of diluted hydrochloric acid added to it, it yields a slight white precipitate of antimonic acid, which dissolves in an excess of the acid. If a drop of the aqueous solution be briskly stirred, on a slip of glass, with a drop of solution of carbonate of soda, it gives a precipitate (antimoniate of soda) which deposits on the lines where the rod has rubbed against the glass.

40. *Examples for Practice in Tables B and C.*—The following substances may be selected (10):—

Corrosive sublimate,	Arsenious acid,
Acetate of lead,	Litharge,
Sulphate of copper,	Red lead,
Oxychloride of bismuth,	Oxide of copper,
Chloride of lead,	Metallic tin.

TABLE D.   51

EXERCISE III. (See (62) for Examples for Practice.)

41. Detection of a Metal belonging to the Sulphide of Ammonium Group.

TABLE D.

EXAMINATION OF THE PRECIPITATE PRODUCED BY SULPHIDE OF AMMONIUM.

This Precipitate may contain—

If Black.			It not Black.*	
Sulphide of Iron, Sulphide of Cobalt, Sulphide of Nickel. Treat a little of the precipitate with HYDROCHLORIC ACID (dil.) (42)			*Sulphide of Manganese* (see 54a), *Phosphate of Lime,*	*Sulphide of Zinc, Oxide of Chromium* (see 59a).

Test separate portions of the original solution with

Dissolved.	Undissolved.		1 AMMONIA		2 POTASH	
Sulphide of Iron. (43.)	Test the original solution with POTASH.		In excess (3). Precipitate, not dissolved by excess; see col. 2. No precipitate, or a precipitate dissolved by excess; add a drop of SULPHIDE OF AMMONIUM. White precipitate indicates *Zinc.* (32, 53.) Pink Precipitate indicates *Manganese.* (54, 55.)	White Precipitate, dissolved by excess of POTASH. *Alumina;* add CHLORIDE OF AMMONIUM, Precipitate *Hydrate of Alumina.* Presence of *Aluminium.*† (30, 31.)	White Precipitate becoming brown when shaken with air. Presence of *Manganese* (54, 55.)	White Precipitate not becoming brown and not soluble in excess. *Phosphate of Lime.* (36, 57, 58, 59.)
Add Potassium Ferrocyanide. Blue Precipitate. Presence of *Iron.* (44, 45.)	Blue Precipitate Presence of *Cobalt.* (46, 47.)	Light green Precipitate Presence of *Nickel.* (48, 49.)				Green Precipitate soluble in excess to a green solution. Presence of *Chromium.* (63 61.)

* This precipitate often exhibits a *green* color, due to the presence of iron in very small quantity, as an impurity. A slight *opaque white* precipitate, increased by the addition of more sulphide of ammonium, is probably accidental, and may be neglected. Should ammonia produce a *bright yellow* precipitate, without adding sulphide of ammonium. it is probably chromate of baryta; see (65).
† Should there be no precipitate, or one which is doubtful, add one or two drops of hydrosulphate of ammonia, to test for zinc, which might possibly have been missed in the adjoining column (52, 53).

NOTES TO TABLE D.

42. When time permits, it is desirable to collect a little of the precipitate upon a filter (4), and to wash it (16) before treating it with hydrochloric acid, which may be poured over it upon the filter.

43. A white or gray residue of sulphur, derived from the sulphide of ammonium, is often left undissolved, especially if the precipitate be treated with hydrochloric acid without being washed.

44. To ascertain whether the iron is present as a ferrous or as a ferric salt, test the original solution, which should be *acid* (18), with ferrocyanide of potassium.

Ferric salts give, with ferrocyanide of potassium, a dark blue precipitate of ferrocyanide of iron (Prussian blue); and with ferridcyanide of potassium, a dark brown solution, but no precipitate.

Ferrous salts give, with ferrocyanide of potassium, a lighter blue precipitate of the double ferrocyanide of iron and potassium; and with ferridcyanide of potassium, a dark blue precipitate of Turnbull's blue.

Potash, added in excess, produces a brown precipitate in solutions of ferric salts, and a dingy green precipitate in those of ferrous salts; this precipitate slowly becomes brown when exposed to the air, from which it absorbs oxygen.

Since nitric acid converts ferrous into ferric salts, no conclusion as to the state of the iron can be drawn from these tests when that acid has been employed in dissolving the substance.

45. *Metallic Iron*, Fe, would generally be recognized by its external appearance. It dissolves slowly in dilute hydrochloric acid, evolving a disagreeable smell of impure hydrogen, quite different from that of hydrosulphuric acid (see *Sulphide of Iron*, p. 55). Ammonia added in excess (5) to this solution, gives a dingy green precipitate which gradually

VARIETIES OF IRON DISTINGUISHED.

Wrought Iron (malleable or bar iron) may be identified by warming two or three grains of it with diluted nitric acid (specific gravity 1.2*) when it will dissolve entirely, yielding a solution which is nearly colorless after cooling, provided that the nitric acid be free from chlorine.

Steel, when tested in this way, gives a solution which retains a brown-yellow color after cooling, due to the products of the action of nitric acid upon the combination of iron with carbon, which is present in steel.

White Cast Iron, heated with nitric acid of the above strength, also dissolves, leaving little or no black residue of uncombined carbon, but the solution has usually a darker color than that furnished by steel, because white iron commonly contains a larger proportion of carbon in combination with the metal.

Gray Cast Iron (ordinary pig iron) leaves a considerable black residue of graphite (uncombined carbon) when heated with nitric acid, and if this be filtered off or allowed to subside, the liquid will usually have a pale brown-yellow color, produced by a little combined carbon.

Mottled Cast Iron furnishes a result intermediate between those obtained with white and gray cast irons, the color of the solution being paler than in the former case, and the black residue less abundant than in the latter.

For the common compounds of Iron, see next page.

---

* Three measures of ordinary strong nitric acid mixed with four measures of water.

Common Compounds of Iron.

Names.	Composition.
Sesquioxide or peroxide of iron, or Ferric oxide	Iron, oxygen.
Magnetic oxide of iron, or Ferroso-ferric oxide	Iron, oxygen.
Sulphate of iron, or Ferrous sulphate	Oxide of iron, sulphuric acid, Water.
Sulphide or sulphuret of iron, or Ferrous sulphide	Iron, sulphur.
Bisulphide of iron, Iron pyrites	Iron, sulphur.
Carbonate of iron, or Spathic iron ore	Oxide of iron, carbonic acid.
Perchloride or sesquichloride of iron	Iron, chlorine.
Ferrocyanide of iron, or Prussian blue	Iron, cyanogen (carbon and nitrogen).
Iodide of iron	Iron, iodine.
Silicilate of iron, or Iron slag	Oxide of iron, silicic acid.

Sesquioxide of Iron, Fe_2O_3, or ferric oxide, is met with in several forms.

Red Hæmatite Ore, or natural sesquioxide of iron, is a hard compact mineral of a dark *reddish-brown* color, not easily reduced to a powder, which is dark red. It is not dissolved by water, but hydrochloric acid slowly dissolves it, yielding a yellow solution, which gives a rust-colored precipitate with ammonia.

Specular Iron Ore, another natural variety of the sesquioxide, is *black*, and has a *brilliant lustre*. Its relations to solvents resemble those of hæmatite.

Brown Hæmatite Ore contains water in combination with ferric oxide, and will therefore give off steam when heated in a dry tube (17). It varies in color through different shades of *yellow, brown, and red*. Hydrochloric acid dissolves it more rapidly than it does the two preceding ores.

The artificial sesquioxide of iron, which is commonly known as *Colcothar, Crocus,* or *Jewellers' Rouge,* has a brighter red color than hæmatite, which it resembles in its behavior with hydrochloric acid.

Rust has the same characters as brown hæmatite.

Magnetic or *Black Oxide of Iron,* Fe_3O_4, as found in nature, is a hard mineral with considerable lustre. Hydrochloric acid slowly dissolves it, with the aid of heat, yielding a greenish solution which gives a dingy green precipitate on addition of ammonia. The black oxide of iron which composes forge scales possesses similar characters, but is usually devoid of lustre.

Sulphate of Iron (copperas, green vitriol, $FeSO_4.H_2O.6Aq.$), forms transparent *green crystals,* often streaked with rusty brown. It dissolves when shaken with cold water, but when boiled with water, it generally deposits a brown basic persulphate of iron formed by the decomposition of some ferric sulphate contained in the salt. This deposit is easily dissolved by hydrochloric acid, forming a yellow solution.

Dried Sulphate of Iron, $FeSO_4$, is a *brownish-white* powder which is not easily dissolved by cold water, and behaves like the green sulphate when boiled. Hydrochloric acid readily dissolves it.

Sulphide of Iron, FeS, is a *black* substance somewhat resembling iron itself, insoluble in water, but dissolving in hydrochloric acid, and evolving a powerful offensive odor of hydrosulphuric acid. On adding ammonia to this solution before the hydrosulphuric acid is boiled off, a black precipitate of sulphide of iron is obtained.

Iron pyrites, FeS_2, has the *color and lustre of pale brass.* It often occurs in distinct cubical crystals, but more commonly in rounded lumps which are dark brown externally, and have a radiated crystalline structure when broken. It is not attacked by water or hydrochloric acid, but nitric acid dissolves it on boiling, with separation of flakes of sulphur

Spathic Iron Ore, $FeCO_3$, or *ferrous carbonate*, varies very much in appearance, but usually forms *grayish-white* masses which seem to be made up of tabular crystals. It is unaffected by water, but dissolves in hydrochloric acid, assisted by heat, with effervescence caused by the escape of carbonic acid.

The *red carbonate of iron* of the druggist is chiefly sesquioxide of iron.

Perchloride of Iron, or *muriate of iron*, or *ferric chloride*, Fe_2Cl_6, is only met with in solution, which has a yellow or red color, and gives a rust-colored precipitate with ammonia. The chlorine may be detected in it by nitrate of silver (Table H). The tincture of sesquichloride of iron is known by its alcoholic smell.

Prussian Blue, $Fe_5C_{12}N_{12}$, is insoluble in water and in the diluted acids. It may be known by its becoming brown when boiled with potash; if the solution of ferrocyanide of potassium thus obtained be filtered from the deposited peroxide of iron, and mixed with excess of hydrochloric acid, it will give a blue precipitate on adding perchloride of iron. Concentrated hydrochloric acid dissolves Prussian blue, on boiling, giving a yellow solution, from which the Prussian blue is precipitated by much water.

Iodide of Iron or *ferrous iodide*, FeI_2, is commonly sold in solution, often mixed with syrup to preserve it. Solution of iodide of iron has a very pale-green color, but, if partly oxidized, it is rusty brown and opaque. It may be tested for iodine according to Table II. The solid iodide forms brownish-green lumps which are deliquescent and have a crystalline fracture.

The *Iron Slags*, or silicates of iron, obtained in the processes of refining and puddling cast iron, are *black* and possessed of considerable lustre. They are unaffected by water, but if finely powdered and boiled with hydrochloric acid (especially if the acid be concentrated), they partly dissolve, emitting an odor of hydrosulphuric acid (caused by the pre-

46. It is desirable to confirm the indication of Cobalt either by the blowpipe (268), or by collecting this precipitate upon a filter (4), washing it (16) and dissolving it off the filter with a little dilute hydrochloric acid. The solution thus obtained should have a light pink color, and if a little of it be dropped upon filter-paper, the latter should become blue or green when gently warmed. Ferridcyanide of potassium added to the solution should give a purple-brown precipitate.

47. Neither metallic cobalt, Co, nor its compounds are frequently met with. The metal bears considerable resemblance to iron.

Cobalt-glance, one of its chief ores, composed of cobalt, arsenic, and sulphur, $CoS_2.CoAs_2$, is a *black lustrous* mineral, soluble in boiling nitric acid, yielding a pink solution, and depositing flakes of sulphur.

Commercial *Oxide of Cobalt*, CoO, is a *bluish-gray, brown* or *black* powder, according to the mode of preparing it. Hydrochloric acid dissolves it to a green or blue liquid which becomes pink when diluted.

Smalt is glass which has been colored with cobalt and powdered. It is insoluble in water, hydrochloric and nitric acids.

Nitrate of Cobalt, $Co2NO_3.6Aq.$, is a red salt, very soluble in water, and easily attracting moisture from the air. It becomes blue when gently heated to expel water of crystallization, and if heated more strongly, evolves brown nitrous fumes, leaving black oxide of cobalt.

48. The indication of nickel may be confirmed by adding to the original solution an excess of ammonia (which gives a blue color, especially on heating), and of a yellow sulphide of ammonium, when the sulphide of nickel first precipitated will, in great measure, be redissolved to a muddy brown solution.

Commercial *Oxide of Nickel*, NiO, is a *dull green* or *brown* powder, which dissolves in hydrochloric acid, yielding a green solution.

Sulphate of Nickel, $NiSO_4.7H_2O$, forms *bright-green crystals*, which are easily dissolved by water to a green solution.

50. The indication of aluminium may be confirmed by observing the character of the precipitate caused by ammonia in the original solution. The hydrate of alumina which is then precipitated is nearly transparent, so that it might easily be overlooked, but that bubbles of air are usually entangled in it. If the liquid be warmed, the alumina separates in more distinct flakes.

The potash employed in testing often contains alumina, which renders it the more necessary to confirm the result by testing the original solution.

In examining substances insoluble in water and acids, silica is likely to be met with here, as well as alumina. To separate them, the original acid solution should be evaporated to dryness (84), and the residue warmed with hydrochloric acid. On pouring the solution into a test-tube, the silica will be seen in flakes, and if these be filtered off, and the solution mixed with excess of ammonia, the flocculent precipitate of alumina will be obtained.

51. *Metallic Aluminium*, Al, resembles tin in appearance, but is much lighter. It is distinguished from all other white metals except platinum, by its resistance to the action of nitric acid, and from platinum by its easily dissolving in hydrochloric acid.

Common Compounds of Aluminium.

Names.	Composition.
Clay or silicate of alumina	Alumina, silicic acid, water.
Sulphate of alumina, or Concentrated alum	Alumina, sulphuric acid, water.
Alum*	Alumina, sulphuric acid, Potash (or ammonia), water.
Emery	Aluminium, oxygen.
Acetate of alumina	Alumina, acetic acid, water.
Aluminate of soda	Alumina, soda.
Kryolite	Aluminium, sodium, fluorine.

Clay, being nearly insoluble in hydrochloric and nitric acids, will be considered hereafter, as will *Emery* and *Kryolite* for the same reason.

Sulphate of Alumina, $Al_2 3SO_4.18Aq.$, is sold as a *white* or *grayish* opaque mass, of crystalline structure, and sweetish astringent taste. It dissolves easily in water, yielding a solution which reddens blue litmus. The sulphuric acid may be detected with chloride of barium (Table II).

Alum may be either sulphate of alumina and potash, $KAl2SO_4.12Aq.$, or sulphate of alumina and ammonia, $NH_4Al2SO_4.12Aq.$, or a mixture of both salts, according to the conditions of its manufacture. It forms bright colorless crystals, which have a sweet astringent taste, and dissolve easily in cold water, yielding a solution which reddens blue litmus. When heated on a knife or a piece of glass alum easily melts, evolves much steam, and leaves a white opaque swollen mass which is quite infusible. If this mass be held so as to touch the outside coating of the flame, the violet-

---

* *Potash-alum* is composed of sulphate of alumina and sulphate of potash, whilst *ammonia-alum* contains sulphate of alumina and sulphate of ammonia.

blue tint imparted to the latter will indicate the presence of potash. Ammonia may be detected by its odor, on boiling the alum with potash.

Acetate of Alumina or *Red Mordant* is sold in a state of solution in water. It has an odor of acetic acid (vinegar), reddens blue litmus, and if diluted with much water and boiled, yields a translucent gelatinous precipitate of basic acetate of alumina. The presence of acetic acid may be proved by the production of a red color on adding perchloride of iron.

Aluminate of Soda, $3Na_2O.Al_2O_3$, is a *grayish-white* opaque substance, strongly alkaline to the taste, dissolving easily in water, yielding a solution which blues red litmus. On adding a single drop of diluted hydrochloric acid, a flocculent precipitate of alumina appears, but is redissolved by an excess of the acid. Addition of ammonia to this solution produces the gelatinous precipitate of alumina.

52. To confirm the presence of zinc, mix the original solution with ammonia in excess, filter, if necessary, and add ferrocyanide of potassium, which produces a white precipitate of ferrocyanide of potassium and zinc, generally appearing yellow through the excess of ferrocyanide of potassium.

Potash produces, in solutions of zinc, a white precipitate which dissolves easily in excess of potash.

(Blowpipe test for Zinc, see 267.)

53. *Metallic Zinc*, Zn, is easily dissolved by hydrochloric acid or nitric acid, the latter serving to distinguish it from aluminium (51).

Common Compounds of Zinc.

Names.	Composition.
Sulphate of zinc, or White vitriol	Oxide of zinc, sulphuric acid, Water.
Oxide of zinc, or Zinc white	Zinc, oxygen.
Carbonate of zinc, or Calamine	Oxide of zinc, carbonic acid.
Sulphide of zinc, or Blende	Zinc, sulphur.
Chloride of zinc	Zinc, chlorine.

Sulphate of Zinc, $ZnSO_4.7Aq.$, is usually met with in *shining needle-like crystals* which dissolve easily in water. The solution reddens blue litmus paper, and has a nauseous metallic taste. The sulphuric acid may be detected by chloride of barium (Table II).

Zinc White, ZnO, is insoluble in water, but dissolves in hydrochloric acid, usually effervescing slightly from the escape of carbonic acid, which the oxide of zinc absorbs from the air. When heated, oxide of zinc becomes yellow, but resumes its white color on cooling.

Carbonate of Zinc, $ZnCO_3$, occurs in nature as *Calamine*, which has a *light brown color* due to the presence of iron. It is insoluble in water, but dissolves with effervescence in hydrochloric or nitric acid. When the nitric solution is mixed with an excess of ammonia, any iron which is present will be precipitated as brown hydrated peroxide, and if this be separated by filtration, the solution will yield a white precipitate with sulphide of ammonium.

Sulphide of Zinc or *Blende*, ZnS, also called *Black Jack*, is another ore of zinc, commonly met with in black shining dodecahedral crystals, the color of which appears to be due to their containing sulphide of iron. It is not affected by water, and dissolves very slowly in boiling hydrochloric acid, evolving the odor of hydrosulphuric acid. Nitric acid dissolves it, generally causing the separation of flakes of sulphur; the solution behaves as described above in the case of calamine.

Chloride of Zinc, $ZnCl_2$, is commonly sold in solution (*Burnett's disinfecting fluid*). The chlorine may be detected by nitrate of silver (Table II). Solid chloride of zinc is white and opaque; it absorbs moisture rapidly from the air, becoming wet (*deliquesces*).

54. To confirm the presence of manganese, boil the original solution with a very little nitric acid* to convert any ferrous salt into ferric salt, and add chloride of ammonium and a slight excess of ammonia (5). Filter from any precipitate caused by iron present as an impurity; test one part of the filtered liquid with potassium ferrocyanide, which should give a white precipitate; to the other part add potassium dichromate, and heat; a dark brown precipitate indicates manganese. (Blowpipe test for Manganese, see 268.)

54*a*. If the manganese be contained in the original substance in the form of manganate or permanganate of potash it will not be detected by the Table, but will be recognized by the dark green or purple-red color of the solution. (See below).

55. *Metallic Manganese*, Mn, is very uncommon. It resembles iron in appearance, but when heated with water it causes effervescence, from escape of hydrogen, accompanied by a very peculiar odor caused by compounds of hydrogen with the carbon contained in the metal.

---

* The addition of nitric acid may be omitted if the original substance was dissolved in hydrochloric acid and evolved the smell of chlorine.

Common Compounds of Manganese.

Names.	Composition.
Binoxide of manganese, or Black manganese	Manganese, oxygen.
Sulphate of manganese	Oxide of manganese, Sulphuric acid, water.
Permanganate of potash	Permanganic acid { manganese, oxygen. } Potash
Manganate of potash	Manganic acid · { manganese, oxygen. } Potash

Binoxide of Manganese, MnO_2, (or *Manganese* as it is often called) is found in nature, as *Pyrolusite*, in *black* shining masses often exhibiting prismatic crystals. The oxide is also met with in dull *dark brown* fragments which give a dark brown powder. It is not attacked by water, but hydrochloric acid dissolves it slowly, evolving a strong smell of chlorine. If the solution be filtered (4) and mixed with ammonia in excess, it often gives a brown precipitate of peroxide of iron, which is a common impurity of the mineral, and after this has been filtered off, the solution will give a buff or flesh-colored precipitate with sulphide of ammonium.

Sulphate of Manganese, $MnSO_4.5H_2O$, is met with (as an artificial product) in crystalline masses of a *pinkish* color, easily dissolved by hot water, yielding a solution which is very nearly colorless. The sulphuric acid may be detected by chloride of barium (Table II).

Permanganate of Potash, $K_2Mn_2O_8$, is commonly sold in the state of solution (*Condy's disinfectant*), known by its magnificent *purple* red color, which is unchanged by cold diluted hydrochloric acid, but vanishes on adding an excess of hydrosulphuric acid, sulphur being separated. If a solution of permanganate of potash be mixed with potash and filtered once or twice through paper, the latter becomes brown from the deposition of binoxide of manganese, and a green solution of manganate of potash passes through.

The solid permanganate of potash forms hard prismatic crystals which appear almost black, but are really dark red with a green reflection, and are at once known by the intense purple color which they impart even to cold water.

Manganate of Potash, K_2MnO_4, is met with as a fine *green* solution which becomes red (permanganate) when diluted with much water, depositing brown hydrated peroxide of manganese. Dilute nitric acid produces a similar change.

56. Several other substances beside phosphate of lime might also be precipitated here, but this being far more common than any of the others, would be found in most cases to compose the precipitate. Phosphate of Alumina (112), Phosphate of Magnesia, Oxalate of Lime, and Fluoride of Calcium are some other common substances which might be met with here. (Phosphate of Baryta and Phosphate of Strontia would also be precipitated, but they are very uncommon.)

In order to be quite sure that phosphate of lime is present, the original solution should be mixed with some acetate of ammonia (prepared by adding acetic acid to ammonia until it reddens blue litmus paper) and divided into two parts.

One part is tested with oxalate of ammonia, which will produce a white precipitate of oxalate of lime.

The other part is tested with *a drop* of perchloride of iron, which will give a white precipitate of phosphate of iron.

The common form of phosphate of lime, $Ca_3 2PO_4$, is *Bone Ash*, which is usually sold as a white powder, insoluble in water but dissolving easily in hydrochloric acid, with slight effervescence, due to the escape of carbonic acid from the carbonate of lime always present in bones, and often leaving a slight dark residue of charcoal.

Superphosphate of Lime, $CaH_4 2PO_4$, is commonly sold as a gray damp powder, which dissolves to a great extent in water yielding a strongly acid solution.

57. *Phosphate of Magnesia* would be known by its not yielding any precipitate on the addition of oxalate of ammonia, as above directed for the detection of lime; but if an excess of ammonia be afterwards added, and the solution briskly stirred (6) a crystalline precipitate of ammonio-phosphate of magnesia will be deposited.

Should there be no precipitate, add some phosphate of soda, and again stir. If this produces a precipitate, it will indicate that magnesia has been precipitated in the wrong place, in consequence of too little chloride of ammonium having been added in Table A.

The only common form of phosphate of magnesia is that of ammonio-phosphate or *triple phosphate* found in calculi, $MgNH_4PO_4$; it is insoluble in water, but dissolves easily in hydrochloric acid.

58. *Oxalate of Lime*, CaC_2O_4, would be precipitated on mixing its original solution with acetate of ammonia, since it is insoluble in the acetic acid thus set free.

Oxalate of lime may be easily identified by its not effervescing when moistened with hydrochloric acid, unless it has been previously heated on a piece of glass or porcelain, when it becomes converted into carbonate of lime, which should be allowed to cool and tested with hydrochloric acid. It may also be tested for oxalic acid according to (114).

59. *Fluoride of Calcium*, CaF_2, or *Fluor Spar* is met with in greenish or purple cubical crystals. Its powder resembles powdered glass in appearance and feel. It is not affected by water, and is dissolved only to a slight extent by diluted hydrochloric acid, the filtered solution yielding a flocculent precipitate with ammonia.

When heated on a knife, or thrown upon a hot surface, fluor spar generally crackles and flies off, at the same time emitting a peculiar phosphorescent light, somewhat resembling that of burning sulphur. It may be tested for fluorine according to Table G.

bichromate of potash, it will not be detected in the Table, but may be recognized by the yellow or red color of the solution, and by the yellow precipitates which it gives with acetate of lead and with nitrate of baryta.

60. The green solution produced by excess of potash will deposit a green precipitate of oxide of chromium when boiled. To confirm the presence of chromium, fuse a very little of the original substance with nitrate of potash, in a small tube (17), when a bright yellow mass of chromate of potash will be produced. This may be dissolved in water, and tested with acetate of lead which gives a yellow precipitate.

61. Metallic chromium, Cr, is not likely to be met with in ordinary analysis.

Common Compounds of Chromium.

Names.	Composition.
Oxide of chromium	Chromium, oxygen.
Chrome iron ore	Chromium, iron, oxygen.
Chromate of potash and Bichromate of potash	Chromic acid (chromium & oxygen) Potash.
Chromate of lead, or Chrome-yellow	Chromic acid, Oxide of lead.
Chrome-alum	Oxide of chromium, sulphuric acid. Potash, water.

Oxide of Chromium, Cr_2O_3, is a *green* powder which is insoluble in water, but generally dissolves, at least partly, in hydrochloric acid, yielding a green solution. If it has been heated to redness, it is insoluble in hydrochloric and in nitric acids, but may be dissolved by boiling it with strong nitric acid, and adding a little chlorate of potash, which oxidizes the chromium into chromic acid, recognizable by the yellow precipitate of chromate of lead which is obtained on adding acetate of ammonia (56) and acetate of lead.

brown or *dark green* color, which is almost insoluble in acids, and may be treated in a similar manner to the preceding, in order to detect the chromium.

Chromate of Potash, K_2CrO_4, forms *bright yellow crystals* easily soluble in water, giving a yellow solution, which becomes orange red when mixed with hydrochloric acid, from the production of bichromate of potash. Hydrosulphuric acid added in large excess to the acidified solution converts it into green chloride of chromium, rendering it opaque from the separation of sulphur.

Bichromate of Potash, $K_2CrO_4.CrO_3$, is sold in large irregular crystals of a fine *orange red* color, easily soluble in water. The solution of this salt also becomes green when mixed with hydrochloric and excess of hydrosulphuric acids.

Chromate of Lead has been described at page 29.

Chrome Alum, $KCr2SO_4.12Aq.$, forms a dark *purple* crystals, which dissolve in water, yielding a purple solution, becoming green when boiled.

62. *Examples for Practice in Table D.*—The following substances may be analyzed (10):—

Chrome alum	Oxide of zinc	Peroxide of iron
Sulphate of nickel	Binoxide of manganese	Sulphate of zinc
Iron pyrites	Common alum	Bone-ash
Sulphate of manganese	Sulphate of iron	Metallic zinc.
Sulphide of iron		

EXERCISE IV. (See (71) for Examples for Practice.)
63. Detection of a Metal belonging to the Carbonate of Ammonia Group.

The Metal may be *Calcium, Barium,* or *Strontium.*

TABLE E.

1.	2.	3.
To a part of the original solution, Add Dil. SULPHURIC ACID. [If no precipitate, pass on to column 3.] Precipitate. See column 2.	To another part of the original solution add SULPHATE OF LIME. Immediate precipitate of *Sulphate of Baryta.* Presence of *Barium* (64, 65). Precipitate formed after a minute or two, or on stirring or warming.* *Sulphate of Strontia.* Presence of *Strontium* (66, 67). No precipitate. See column 3.	To a fresh part of the original solution add AMMONIA in excess (5) and OXALATE OF AMMONIA. Precipitate *Oxalate of Lime.* Presence of *Calcium*† (68, 69).

* If the substance has been dissolved in an acid, the original solution must be evaporated to dryness (84), and the residue dissolved in water before testing for strontium with sulphate of lime.

† If neither barium, strontium, nor calcium can be detected, it is probable that the carbonate of ammonia had precipitated the magnesium, as sometimes happens on boiling, if too little chloride of ammonium has been added. Test for magnesium, by adding phosphate of soda to the solution which has been tested for calcium in col. 3.

NOTES TO TABLE E.

64. To confirm the presence of barium, add hydrofluosilicic acid to a portion of the original solution, and stir well with a glass rod (6), when a semi-transparent crystalline precipitate of silicofluoride of barium will be deposited.

Or, dissolve the carbonate of ammonia precipitate in acetic acid, and add chromate of potash, which produces a yellow precipitate of chromate of baryta.

65. *Metallic Barium*, Ba, is very uncommon. It speedily oxidizes when exposed to the air, becoming converted into baryta.

COMMON COMPOUNDS OF BARIUM.

Names.	*Composition.*
Sulphate of baryta, or Heavy spar	Baryta (oxide of barium), sulphuric acid.
Carbonate of baryta, or Witherite	Baryta, carbonic acid.
Chloride of barium	Barium, chlorine, water.
Nitrate of baryta	Baryta, nitric acid.
Hydrate of baryta	Baryta, water.
Baryta	Barium, oxygen.
Chlorate of baryta	Baryta, chloric acid.
Chromate of baryta	Baryta, chromic acid.

Sulphate of Baryta, $BaSO_4$, being insoluble in water and acids, will be considered hereafter.

Carbonate of Baryta, $BaCO_3$, is found as a *grayish-white*, heavy earthy mineral, insoluble in water, but dissolving, with effervescence, in hydrochloric acid. If a glass rod or a platinum wire be dipped into the solution, and held in the margin of a flame (70), it will tinge the flame green. The artificial carbonate of baryta is a *pure white* earthy powder, which usually gives a milky solution in hydrochloric acid, from a slight impurity of sulphate of baryta, not easily filtered off.

Chloride of Barium or *muriate of barytes*, $BaCl_2.2Aq.$, forms *transparent flat crystals*, which dissolve easily in water. The chlorine may be detected by nitrate of silver (Table H).

Nitrate of Baryta, $Ba2NO_3$, forms *colorless crystals*, which are dissolved by water. When heated in a dry tube (17), it evolves brown fumes of nitric peroxide, and leaves an infusible residue of baryta.

Hydrate of Baryta, BaH_2O_2, is sold either in colorless crystals or as a white powder. The crystals soon become opaque when exposed to the air, losing water and absorbing carbonic acid. Hydrate of baryta dissolves easily in water, yielding a strongly alkaline solution, soon rendered milky by the carbonic acid of the air, especially if shaken in a test-tube which has been breathed into.

Baryta, BaO, is sold as a gray porous solid, which becomes very hot when moistened with water, and crumbles to a white powder of hydrate of baryta.

Chlorate of Baryta, $Ba2ClO_3$, forms colorless crystals soluble in water. Hydrochloric acid colors the solution yellow, on applying heat, by decomposing the chloric acid, and evolves a chlorous smell. Hydrosulphuric acid added to this solution yields a white deposit of sulphur. Strong sulphuric acid reddens the chlorate, and causes explosion on heating.

Chromate of Baryta, $BaCrO_4$, is a bright yellow powder, insoluble in water, but soluble in hydrochloric acid, giving a yellow solution, which becomes green when heated with alcohol or allowed to remain in contact with metallic zinc, chloride of chromium being formed.

66. To confirm the presence of strontium, apply the colored flame test (70) to the original substance.

67. *Metallic Strontium*, Sr, is a rarity; it is soon converted into strontia by absorbing oxygen from the air.

Common Compounds of Strontium.

Names.	Composition.
Nitrate of strontia	Strontia (oxide of strontium), Nitric acid, Water.
Sulphate of strontia, or Celestine	Strontia, sulphuric acid.
Carbonate of strontia, or Strontianite	Strontia, carbonic acid.

Nitrate of Strontia, $Sr2NO_3.5Aq.$, forms *colorless crystals*, which are easily dissolved by water; paper dipped into the solution and held in the margin of a flame (70) will color it with flashes of crimson.*

When heated in a dry tube (17), nitrate of strontia evolves water and brown fumes of nitric peroxide, leaving an infusible residue of strontia. Nitrate of strontia which has been crystallized by boiling down its solution contains no water.

Sulphate of Strontia, $SrSO_4$, being nearly insoluble in water and acids, will not be considered here.

Carbonate of Strontia, $SrCO_3$, is usually found as a *greenish* mineral, which is insoluble in water, but dissolves in hydrochloric acid with effervescence. Paper dipped into the solution colors flame crimson* (70).

68. Oxalate of Ammonia also forms precipitates in solutions containing barium and strontium, so that it can never be used as a test for calcium, unless the absence of those metals has been previously ascertained.

69. *Metallic Calcium*, Ca, is not likely to be met with

---

* Some care is requisite to avoid mistaking the orange-red tint which calcium imparts to flame, for the crimson of strontium.

in ordinary analysis; it quickly absorbs oxygen from the air, and is converted into lime.

Common Compounds of Calcium.

Names.	Composition.
Lime, or Quick lime	Calcium, oxygen.
Hydrate of lime, or Slaked lime	Lime, water.
Carbonate of lime, or Chalk	Lime, carbonic acid.
Sulphate of lime, or Gypsum	Lime, sulphuric acid, water.
Chloride of calcium	Calcium, chlorine.
Chloride of lime, or Bleaching powder, or Hypochlorite of lime	Lime, hypochlorous acid, Chloride of calcium, Water.
Oxalate of lime	Lime, oxalic acid.
Phosphate of lime	Lime, phosphoric acid.
Superphosphate of lime	Lime, phosphoric acid, water.
Fluoride of calcium	Calcium, fluorine.

Quick Lime, CaO, is a *grayish white* earthy solid, which becomes hot when moistened with water, and crumbles after a time to a white powder of hydrate of lime.

Hydrate of Lime, CaH_2O_2, is a light *white* powder, which is not visibly dissolved by water, but dissolves easily in hydrochloric acid, usually effervescing slightly, from the presence of a little carbonate of lime. When shaken with cold water, and filtered, hydrate of lime gives a solution which turns red litmus blue, and becomes milky when shaken in a test-tube which has been breathed into, from the precipitation of carbonate of lime.

Lime-water may be recognized by the test just given.

Carbonate of Lime, $CaCO_3$, occurs in nature in several forms, all of which, however, are insoluble in water, but dissolve easily in hydrochloric acid, with brisk effervescence from escape of carbonic acid.

Limestone and Chalk, which are the most impure of the natural varieties of carbonate of lime, generally yield a slight flocculent precipitate of alumina when their hydrochloric solution is tested with ammonia, and a green tinge, from the presence of iron, when sulphide of ammonium is added to the ammoniacal liquid. *Marble Iceland Spar*, and prepared chalk, which are purer forms of the carbonate, do not behave in the same way.

Magnesium Limestone, or *Dolomite*, $CaMg2CO_3$, which contains the carbonates of lime and magnesia, may be identified by dissolving it in diluted hydrochloric acid, adding chloride of ammonium, ammonia, and carbonate of ammonia, to precipitate the lime, boiling, filtering, and testing the solution for magnesia with phosphate of soda, after proving, by the addition of oxalate of ammonia, that all the lime has been separated.

Sulphate of Lime, $CaSO_4$, does not visibly dissolve in water, although if the solution be filtered, a small quantity of the salt will be found in solution. The sulphuric acid may be detected by chloride of barium (Table II). Hydrochloric acid dissolves it to a greater extent, but, unless after prolonged boiling, it might easily be concluded that sulphate of lime was not dissolved by water or acids, so that it is often found among substances of that class (Table I).

The several varieties of sulphate of lime differ considerably in appearance.

Gypsum, $CaSO_4.2H_2O$, is a grayish-white, opaque, earthy, brittle mineral.

Fibrous gypsum is made up of parallel silky fibres, which are white, gray, or pink.

Selenite is transparent, or nearly so, either colorless, or brownish-gray in color, and easily split into plates with a

which sets into a solid mass after a few minutes, if mixed with water to a thin paste.

Chloride of Calcium is sold in three forms. The crystallized chloride, $CaCl_2.6Aq.$, forms colorless transparent crystals which rapidly absorb water from the air, and are extremely soluble in water. Another variety is the porous chloride, forming a white or grayish-white porous mass, $CaCl_2.2Aq.$, which becomes wet (deliquesces) very rapidly when exposed to air, and dissolves easily in water. The fused chloride, $CaCl_2$, has similar properties, but is a gray fibrous crystalline solid.

The chlorine may be detected by nitrate of silver. (Table H.)

Chloride of Lime, or bleaching powder, $2CaHClO_2.CaCl_2. 2H_2O$, is a white earthy powder, which has a strong smell of hypochloric acid, resembling that of chlorine. It is partly dissolved by water, and entirely by hydrochloric acid, with effervescence, evolving a powerful odor of chlorine. Litmus paper is at once bleached by the solution.

Oxalate of Lime, CaC_2O_4, in a pure state, is a white powder, insoluble in water, but soluble in hydrochloric acid. The method of identifying it has been described at (58).

Phosphate and Superphospate of Lime have also been described at (56).

Fluoride of Calcium will be found described at (59).

70. *Colored Flame Test.*—Although this test will be more fully described in the Exercises with the Blowpipe, it is necessary to refer to it here, because it affords so valuable a confirmation of the results obtained by liquid tests in Table E.

The chlorides of the metals furnish the most distinct colored flames, since they are more easily vaporized and mingled with the gases of the flame, when the hydrogen abstracts the chlorine, and the metallic vapor burns with its characteristic tint. Hence the substance to be tested should be dissolved, if possible, in hydrochloric acid.

COLORED FLAME TEST.

The best flame for this test is that of a Bunsen's air-burner (fig. 23); but if this be not at hand, a fair substi-

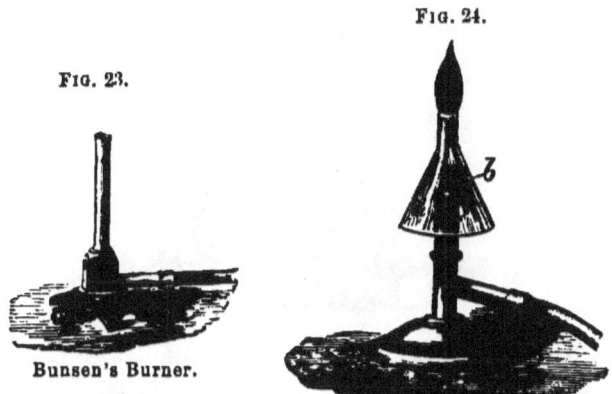

Fig. 23.

Bunsen's Burner.

Fig. 24.

tute may be made by placing a small glass funnel (*a*, fig. 24), with a rather wide neck, over a gas-burner, as shown in fig. 24, where (*b*) is one of the flattened burners often employed for blowpipe experiments. The funnel is placed over the jet before the gas is lighted; the gas is then turned on to a moderate extent, when it mixes with the air passing up through the funnel, and by gradually diminishing the supply of gas, it may be made to burn with a nearly non-luminous flame, in the margin of which the glass rod, or, better, platinum wire (74), moistened with the liquid under examination, should be held.

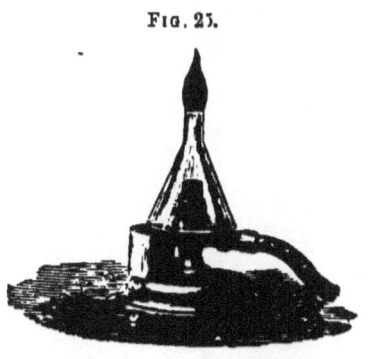

Fig. 25.

be supported, so that a passage for air may be left between the glass and the jet.

Fig. 26.

A spirit-lamp (fig. 26) may be used when gas is not attainable, though it does not give so good results as the gas-flame.

71. *Examples for Practice in Table E.*—The following compounds may serve as exercises in this Table (10):—

Chloride of barium
Nitrate of strontia
Carbonate of baryta

Chloride of calcium
Carbonate of lime (chalk or marble)
Sulphate of lime (plaster of Paris).

EXERCISE V. (See (81) for Examples for Practice.)

72. Detection of Potassium, Sodium, and Ammonium.

TABLE F.

| Boil a portion of the original substance with POTASH. Odor of *Ammonia* indicates the presence of *Ammonium* (75). | Make a rather strong solution of the substance, and stir it (73) with TARTARIC ACID,* or with BICHLORIDE OF PLATINUM on a slip of glass (76). Precipitate formed in lines. Presence of *Potassium* (77). | Take a little of the substance upon a PLATINUM WIRE (74), and expose in flame (70) strong yellow color (78). Presence of *Sodium* (80). Or, make a rather strong solution (79), and stir with ANTIMONIATE OF POTASH† on a slip of glass (73). Precipitate in lines. Presence of *Sodium* (80). |

* If the original solution be alkaline (18), enough tartaric acid must be added to render it distinctly acid ; or it be tested with platinum chloride, it must be previously acidified with hydrochloric acid. Acid solutions must rendered slightly alkaline with sodium carbonate before testing with tartaric acid.
† If the solution is acid it must be rendered alkaline by potash before testing.

EXPLANATIONS AND INSTRUCTIONS ON TABLE F.

73. The precipitation of potassium with tartaric acid or bichloride of platinum, and of sodium with antimoniate of potash, is more readily effected by stirring on a slip of glass than in a test tube.

Take a clean, dry slip of window-glass; dip a glass rod (6) into the solution to be tested, and place the drop so withdrawn upon the slip of glass. Wipe the rod clean, and dip it into the test, placing the drop withdrawn by the side of the other; notice that both drops are clear, and stir them briskly together with the end of the glass rod, which should be moved in circles, but not hard enough to scratch the glass (fig. 27). The precipitate will then be deposited in lines (fig. 28) upon those parts of the slip of glass which have been rubbed by the rod.

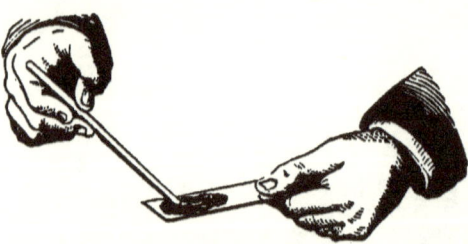

Fig. 27.

Fig. 28.

74. The platinum wire for this purpose should be very thin, so that one inch may weigh $\frac{1}{4}$ grain. A piece about three inches long should be fixed into a glass handle, which is conveniently made by softening the centre of a narrow glass tube in the blowpipe-flame (fig. 15), drawing it out to

Fig. 29.

a narrow neck (fig. 19), and cutting it off at *a*; the platinum wire is then inserted (fig. 29), and the glass fused round it

Since all platinum wire which has been fingered tinges flame yellow (sodium having been derived from the perspiration of the skin), it must be cleansed before use by holding it in the margin of the flame until a yellow tinge is no longer visible.

NOTES TO TABLE F.

75. *Ammonium*, NH_4, is not known to have any separate existence, but it is often very convenient to represent the nitrogen and hydrogen in the salts of ammonia, as existing in the form of a compound metal, capable of taking the same part in the composition of those salts as is taken by potassium and sodium in their salts.

COMMON COMPOUNDS OF AMMONIA.

Names.	Composition.
Solution of ammonia, or *Liquor ammoniæ*	Ammonia, { Nitrogen, Hydrogen. } Water.
Carbonate of ammonia	Ammonia, water, carbonic acid.
Sulphate of ammonia	Ammonia, water, sulphuric acid.
Nitrate of ammonia	Ammonia, water, nitric acid.
Sulphide of ammonium	Ammonia, hydrosulphuric acid.
Chloride of ammonium, or Sal-ammoniac	Ammonia, hydrochloric acid.
Oxalate of ammonia	Ammonia, water, oxalic acid.

Solution of Ammonia, NH_3, has the strong odor of *hartshorn*, does not effervesce with dilute hydrochloric acid, and leaves no residue when evaporated on a slip of glass.

Chloride of Ammonium, or *muriate of ammonia*, or *hydrochlorate of ammonia*, or *sal-ammoniac*, NH_4Cl, is sold either in white crystals or in translucent fibrous masses, usually stained brown in places. It is very easily dissolved by water. It has no ammoniacal smell, and when heated on a knife or a slip of glass, it evaporates in white fumes without melting. The chlorine may be detected with nitrate of silver. (Table

Carbonate of ammonia, or *sesquicarbonate of ammonia*, or *Preston salts*, $2(NH_4)_2CO_3.CO_2$, has a powerful odor of ammonia. It is usually sold in white opaque lumps, which are transparent when freshly prepared. Carbonate of ammonia dissolves easily in water, yielding a solution which blues red litmus paper, and effervesces violently with hydrochloric acid, in consequence of the escape of carbonic acid.

Sulphate of ammonia, $(NH_4)_2SO_4$, forms prismatic crystals, which are colorless when pure, but in their impure state have a brownish color. It dissolves easily in water, and has no ammoniacal odor. The sulphuric acid may be detected by chloride of barium. (Table H.)

Nitrate of Ammonia, NH_4NO_3, is sold either in colorless crystals or in opaque fused masses. It does not smell of ammonia, and becomes damp on exposure to air; easily soluble in water. When heated on a slip of glass, it melts very easily, boils, and passes off entirely as nitrous oxide gas and steam. The nitric acid may be detected as in Table H.

Sulphide of Ammonium, or *hydrosulphate of ammonia*, $(NH_4)_2S$, is common in a state of solution only.

The solution is *yellow* (though colorless when quite freshly prepared), and has a very offensive ammoniacal smell, and an alkaline reaction. The addition of hydrochloric acid causes a milkiness, due to the precipitation of sulphur, and an escape of hydrosulphuric acid, recognized by its odor.

Oxalate of Ammonia, $(NH_4)_2C_2O_4.Aq.$, forms shining, white, needle-like crystals, which are free from ammoniacal smell, and dissolve easily in water. The oxalic acid may be detected according to Table II.

76. The precipitate produced by tartaric acid is the bitartrate of potash, $KHC_4H_4O_6$. The (yellow) precipitate produced by platinum chloride is the platinochloride of potassium, $2KCl.PtCl_4$. Since ammonium is precipitated by the same tests, it is absolutely necessary to prove its absence

Should platinum chloride produce a *dark red* color, iodine is probably present. (See *Iodide of Potassium* (77).)

77. *Metallic Potassium*, K, is not met with in ordinary analysis. It is oxidized immediately by exposure to air, and takes fire in contact with water, burning with a violet flame.

Common Compounds of Potassium.

Names.	Composition.
Nitrate of potash, or Saltpetre	Potash { Potassium, Oxygen. } Nitric acid.
Carbonate of potash, or Pearlash	Potash, carbonic acid.
Bicarbonate of potash	Potash, carbonic acid, water.
Sulphate of potash	Potash, sulphuric acid.
Bisulphate of potash	Potash, sulphuric acid, water.
Chloride of potassium	Potassium, chlorine.
Hydrate of potash	Potash, water.
Bitartrate of potash	Potash, tartaric acid, water.
Chromate of potash, Bichromate of potash	Potash, chromic acid.
Chlorate of potash	Potash, chloric acid.
Ferrocyanide of potassium, or Prussiate of potash	Potassium, cyanogen { Carbon, Nitrogen. } Iron, water.
Ferridcyanide of potassium, or Red prussiate of potash	Potassium, cyanogen, iron.
Cyanide of potassium	Potassium, cyanogen { Carbon, Nitrogen. }
Iodide of potassium	Potassium, iodine.
Acid oxalate of potash	Potash, oxalic acid, water.
Silicate of potash	Potash, silicic acid.
Soft soap	Potash, oleic acid, water.

Nitre, or nitrate of potash, KNO_3, is easily soluble in water, and readily deposits in prismatic crystals from a hot and strong solution allowed to cool.

Placed on the point of a knife and held in the margin of a flame (70), it melts, boils, and colors the flame blue violet.

Heated in a dry tube (17), it easily melts to a clear liquid, in which a piece of wood or paper burns with vivid *deflagration*. Brown nitrous fumes may afterwards be seen and smelt in the upper part of the tube.

The nitric acid may be detected according to Table II.

Nitre is met with in commerce in several forms.

The *grough*, or impure nitre, as imported from India, consists of small brownish irregular crystals, owing its color to the presence of vegetable matter from the earth out of which it is extracted.

The common saltpetre of the shops forms colorless irregular crystalline lumps.

Refined saltpetre forms colorless prismatic crystals, often marked with longitudinal grooves, or else a pure white crystalline powder (*saltpetre flour*).

Sal Prunelle is saltpetre which has been melted and cast into the form of opaque white bullets.

Chlorate of Potash, $KClO_3$, forms colorless flat crystals. It is not easily soluble in cold water. Hot water dissolves it, but readily deposits it in flat crystals on cooling. The solution of chlorate of potash becomes yellowish and emits a chlorous odor when heated with hydrochloric acid, the chloric acid being decomposed, and when hydrosulphuric acid is added to the acidified solution, a white milky precipitate of sulphur is obtained.

Heated in a dry tube (17), it easily melts to a clear liquid, which soon boils and evolves oxygen, recognized by its kindling into a blaze a spark at the end of a match held at the mouth of the tube.

Carbonate of Potash, K_2CO_3, soon becomes damp (deliquesces) when exposed to air, from absorption of water. It dissolves easily in cold water, yielding a very alkaline solution which effervesces briskly on adding hydrochloric acid.

The carbonate of potash which is known as *American potash*, or *pearlash*, forms bluish-white half-fused lumps. *Salt*

Bicarbonate of Potash, $KHCO_3$, is sold either in transparent prismatic crystals, or as a white powder. It does not deliquesce in air, dissolves less easily in cold water than the carbonate, and the solution is not so strongly alkaline.

When the solution of bicarbonate of potash is heated to boiling it effervesces slowly, from the escape of carbonic acid. If solution of sulphate of magnesia be added to a solution of bicarbonate of potash (prepared with cold water) it does not produce a precipitate until the solution is boiled, whilst the carbonate of potash produces a precipitate without boiling.

Sulphate of Potash, K_2SO_4, forms hard, colorless, prismatic crystals, which do not dissolve very quickly in cold water. Its solution does not redden blue litmus paper.

Bisulphate of Potash, $KHSO_4$, dissolves more easily in water, yielding a strongly acid solution, reddening blue litmus paper.

Chloride of Potassium, KCl, forms white cubical crystals, which crackle (decrepitate) when heated on a slip of glass, and dissolve very easily in cold water. The chlorine may be detected by nitrate of silver (Table H).

Hydrate of Potash, KHO (*caustic potash*, or *potassa fusa*), is sold either in lumps or round sticks somewhat resembling porcelain, and generally cream-colored (older samples have a blue color). It becomes wet almost immediately when exposed to air, and dissolves very quickly in cold water, producing much heat, and a strongly alkaline solution.

The *liquor potassæ*, or solution of potash, generally contains a little carbonate of potash, but it may be distinguished from a solution of that salt by its not effervescing with the first drop or two of hydrochloric acid, the carbonic acid not escaping until the whole of the hydrate of potash has been neutralized. Mercuric chloride gives a bright yellow precipitate with the solution of potash. Nitrate of silver gives a dark brown precipitate.

soluble with some difficulty in cold water, but dissolves in boiling water, and is deposited in shining crystals on cooling. The solution reddens blue litmus paper. Hydrochloric acid easily dissolves it. When heated on a slip of glass, or the blade of a knife, bitartrate of potash blackens, from the separation of charcoal, and emits a peculiar odor of burnt sugar, due to the decomposition of the tartaric acid. The residue, when moistened with water, turns red litmus blue, and effervesces strongly with hydrochloric acid, the bitartrate of potash having been converted into carbonate.

Cream of tartar is a white crystalline powder.

The impure Bitartrate of Potash known as *Argol*, is sold in irregular crystalline lumps of a brown or dark purple color, derived from the grape-juice which deposits it. Such Argol often contains much tartrate of lime. Refined Argol forms white crystalline lumps.

Chromate and *Bichromate of Potash* were described at (61).

Ferrocyanide of Potassium, $K_4C_6N_6Fe.3Aq.$, is sold in *yellow* crystalline masses. It dissolves easily in water to a yellow solution, which becomes blue when mixed with hydrochloric acid and warmed, at the same time evolving the peculiar odor of prussic acid. Perchloride of iron gives a dark-blue precipitate of Prussian blue with the solution of ferrocyanide of potassium.

Ferridcyanide, or *Ferricyanide of Potassium*, $K_3C_6N_6Fe$, forms dark-red prismatic crystals, which dissolve easily in water, giving a green solution. The solution gives an intensely blue precipitate with sulphate of iron.

Cyanide of Potassium, KCN, is sold in white porcelain-like masses or sticks, which smell of prussic acid, and slightly of ammonia, resulting from decomposition. The cyanide soon becomes damp in air, and dissolves very easily in cold water. The solution is strongly alkaline. Commercial cyanide of potassium always contains carbonate and cyanate of potash, so that it effervesces strongly on addition of hydrochloric acid, evolving a powerful odor of prussic acid.

The cyanogen may be detected according to (98).

Iodide of Potassium, or *hydriodate of potash*, KI, forms white cubical crystals, often brownish after exposure to the air of the laboratory, from separation of a little iodine. It dissolves very easily in cold water. If the solution be boiled with a little nitric acid, it evolves violet vapors of iodine.

Acid Oxalate of Potash, or *salt of sorrel*, or *essential salt of lemons*, may be either the binoxalate, $KHC_2O_4.Aq$., or quadroxalate of potash, $KH_3 2C_2O_4.2Aq$. It forms hard white crystals, not easily dissolved by cold water, but soluble in hot water, yielding a solution which strongly reddens blue litmus. When heated on a knife-blade or a slip of glass, the oxalate (is not blackened, like the bitartrate, but) is converted into carbonate of potash, which may be recognized by its strongly bluing moistened red litmus, and effervescing when moistened with hydrochloric acid.

Silicate of Potash, K_4SiO_4, or *Soluble Glass*, is sold either as a gummy liquid or in fused masses, which dissolve slowly in water. The solution of silicate of potash is strongly alkaline; when diluted hydrochloric acid is gradually added to it, slight effervescence generally takes place, from the presence of a little carbonate of potash, and when the solution is nearly neutralized, the silicic acid begins to separate in the gelatinous form, especially on heating, sometimes converting the whole solution into a jelly. The presence of silicic acid may be established beyond doubt according to (118).

Soft Soap, or *Oleate of Potash*, $KC_{18}H_{33}O_2$, is known by its peculiar appearance and smell. It dissolves in water, yielding an alkaline solution, which becomes milky with diluted hydrochloric acid, from the separation of oleic acid. On boiling the acidified solution, the oleic acid collects on the surface as an oily layer.

78. Since a very minute quantity of sodium will impart a distinct yellow color to flame, it often happens that a little of this substance present as an impurity is regarded by the

stance under examination. To avoid error, some collateral evidence must be sought for. Thus, it should be ascertained whether the substance under examination possesses the characters of any of the compounds of sodium described (in 80).

If the original solution be neutral or alkaline to test-paper, and no metal has been found by the application of any previous tests, then it may be inferred, if the substance imparts a distinct bright yellow color to the flame, that sodium is an essential constituent of it.

79. Although antimoniate of potash is a very excellent test for sodium, when the solution is freshly prepared, it does not answer so well in dilute solutions containing sodium, if the antimoniate has been kept for some time in solution. Another objection to the test is the circumstance that very small quantities of lime, and some other bases, will give bulky precipitates with the antimoniate of potash, altogether misleading the analyst. Free acids also cause a milky precipitate of antimonic acid.

80. *Metallic Sodium*, Na, is a soft metal, with a silvery lustre when freshly cut, but tarnishing with extreme rapidity when exposed to air. Thrown upon water, it fuses, and the silvery globule floats over the surface, emitting a hissing sound, from the escape of hydrogen; on applying a light, the hydrogen burns with a bright yellow flame.

For the common compounds of sodium, see next page.

Common Compounds of Sodium.

Names.	Composition.
Carbonate of soda	{ Soda (oxide of sodium). { Carbonic acid.
Bicarbonate of soda	Soda, carbonic acid, water.
Hydrate of soda, or Caustic soda	Soda, water.
Soda-ash	{ Hydrate of soda. { Carbonate of soda.
Common salt	Sodium, chlorine.
Sulphate of soda	Soda, sulphuric acid.
Nitrate of soda	Soda, nitric acid.
Sulphite of soda	Soda, sulphurous acid, water.
Hyposulphite of soda	Soda, hyposulphurous acid, water.
Chloride of soda	{ Hypochlorite of soda. { Chloride of sodium, water.
Phosphate of soda	Soda, phosphoric acid, water.
Arseniate of soda	Soda, arsenic acid, water.
Biborate of soda, or Borax	Soda, boracic acid, water.
Silicate of soda	Soda, silicic acid.
Tungstate of soda	Soda, tungstic acid.
Soda-soap, or Hard soap	{ Soda, stearic, oleic, or palmitic { acid, water.

Carbonate of soda, Na_2CO_3, dissolves easily in water, yielding a strongly alkaline solution, which effervesces strongly, from escape of carbonic acid, when hydrochloric acid is added.

Common *washing-soda*, $Na_2CO_3.10Aq.$, is crystallized carbonate of soda, containing nearly two-thirds of its weight of water. The crystals *effloresce*, or become opaque at the surface when exposed to the air, from loss of water. When heated on a knife or a slip of glass, crystallized carbonate of soda melts, boils, evolves much steam, and leaves a white dry residue, which requires a blowpipe-heat to fuse it.

The carbonate of soda in powder, which is sold by the druggist is a *bicarbonate of soda*, $NaHCO_3$, which is less easily dissolved by water than the true carbonate; the solution is not so strongly alkaline, and effervesces when boiled, from escape of carbonic acid. It may also be distinguished from the true carbonate by testing it with sulphate of magnesia. See *Bicarbonate of Potash*.

Hydrate of Soda, NaHO, or *Caustic Soda* is commonly sold in opaque white fused masses which rapidly absorb moisture from the air. It dissolves very easily in water, evolving heat, and yielding a very strongly alkaline solution which effervesces very slightly, if at all, with hydrochloric acid.

Soda-Ash is a mixture of carbonate of soda and hydrate of soda, which has the appearance of earthy lumps or coarse powder. It dissolves in water, generally leaving a slight flaky residue of impurities, including some particles of carbonaceous matter. Its solution does not effervesce on the addition of the first two or three drops of hydrochloric acid, these being neutralized by the hydrate of soda, but a further addition of the acid decomposes the carbonate of soda, with effervescence. The solution generally contains traces of alumina and lime.

Common Salt or *Chloride of Sodium*, NaCl, is of course easily recognized by its taste. It is readily soluble in cold water. The chlorine may be detected by nitrate of silver (Table H).

Sulphate of Soda or *Glauber's Salt*, $Na_2SO_4.10Aq.$, is usually sold in transparent prismatic crystals, which soon become opaque (*effloresce*) when exposed to the air, from loss of water of crystallization. It dissolves easily in water. The sulphuric acid may be detected with chloride of barium (Table H).

Salt-cake is fused sulphate of soda, Na_2SO_4, and forms opaque white masses which are much less pure than the crystalline sulphate, so that they do not give a clear solution in water, and the solution is acid, from the presence of some bisulphate of soda.

Nitrate of Soda, $NaNO_3$, or *Peruvian* or *Chili saltpetre* or *cubic nitre*, forms colorless crystals; the crude salt, however, is often brown or gray. It becomes moist when exposed to air, and dissolves very easily in water. Placed on the point of a knife, and held in the margin of a flame, it colors it intensely yellow. When heated in a dry tube, it behaves like nitrate of potash (77). The nitric acid may be detected according to Table II.

Sulphite of Soda, $Na_2SO_3.7Aq.$, when freshly prepared, forms transparent crystals, but they soon become opaque at the surface when exposed to air. It dissolves easily in water, yielding a solution which turns red litmus paper blue, and has a decidedly sulphurous taste. On adding diluted hydrochloric acid, the solution evolves the odor of sulphurous acid, and hydrosulphuric acid renders the acidified solution milky by causing the separation of sulphur.

Hyposulphite of Soda, $Na_2S_2O_3.5Aq.$, forms brilliant transparent crystals which dissolve very easily in water. On adding diluted hydrochloric acid to the solution, it slowly becomes milky and acquires a yellow color, from the separation of sulphur, the odor of sulphurous acid being perceptible at the mouth of the tube.

Hypochlorite of Soda, $NaClO$, is always sold in solution under the names of *Chloride of Soda* and *Liquor Sodæ Chlorinatæ*. It has a strong smell of hypochlorous acid (somewhat resembling that of chlorine). On adding diluted hydrochloric acid, it becomes yellowish and evolves a powerful odor of chlorine. Test-papers are at once bleached by the acidified liquid.

Phosphate of Soda (common phosphate, orthophosphate, or rhombic phosphate of soda, $Na_2HPO_4.12Aq.$) forms transparent crystals, which effloresce, or become opaque from loss of water, when exposed to the air. It dissolves easily in water, and the solution turns red litmus paper blue. The phosphoric acid may be detected according to Table H.

Arseniate of Soda has been described at (35).

Biborate of Soda or *Borax*, $Na_2O.2B_2O_3.10Aq.$, is commonly sold either as a white powder, or in transparent crystals which lose water and become opaque when exposed to air. It dissolves easily in water, and the solution turns red litmus paper blue. When heated on a knife or a slip of glass, it melts and swells up, evolving steam, and leaving a white porous mass. When fused in the blowpipe-flame in a loop of platinum wire (243) it forms a bead of glass which remains transparent on cooling. The boracic acid may be detected according to Table H.

Glass of Borax or vitrified borax, $Na_2O.2B_2O_3$, forms transparent or semi-transparent glassy-masses, which are dissolved slowly even by boiling water, and require the blowpipe-flame to effect their fusion.

Silicate of Soda, Na_4SiO_4, or *Soluble Glass* is generally sold as a grayish gummy solution, which behaves in the same manner as the solution of silicate of potash (p. 85).

Tungstate of Soda, $Na_2WO_4.2Aq.$, is commonly sold in opaque, irregular crystals, which dissolve easily in water, yielding an alkaline solution, which gives a white precipitate of tungstic acid on adding diluted hydrochloric acid; if a piece of zinc be placed in the solution to which an excess of hydrochloric acid has been added, a beautiful blue oxide of tungsten is gradually formed.

Soda-soap is soluble in warm water, giving a solution which turns red litmus paper blue, and gives a white precipitate of stearic acid on addition of hydrochloric acid. When this precipitate is boiled in the liquid, it collects as an oily layer upon the surface.

81. *Examples for Practice in Table* **F.**—The following substances may be analyzed for practice (10):—

Chloride of ammonium	Carbonate of soda
Bicarbonate of potash	Carbonate of potash
Chloride of sodium	Sulphate of soda
Biborate of soda	Nitrate of potash.

EXERCISE VI. (See (121) for Examples for Practice.)

82. Analysis of Solid Substances containing a Single Acid or Non-metal

(the Organic Acids being left out of consideration for the present).

TABLE G.

EXAMINATION OF SOLID SUBSTANCES CONTAINING ONE ACID OR NON-METAL (NOT BEING ORGANIC) (85).

1.	2.	3.
Treat the powder with STRONG SULPHURIC ACID (86).	Heat another part with DIL. HYDROCHLORIC ACID.	Dissolve (3) a fresh portion in WATER or HYDROCHLORIC or NITRIC ACID.†
Effervescence indicates *Carbonic Acid*; nearly inodorous. See col. 2. *Hydrochloric Acid*; pungent clouds (87). See col. 3. Yellow or red color; possibly *Chloric Acid* (90).	*a.* Effervescence indicates *Carbonic Acid*; nearly inodorous (94, 95). *Hydrosulphuric Acid*; putrid odor (96, 97).	Filter, if necessary, and examine as in Table H.
Heat moderately* *a. Hydrofluoric Acid*; pungent clouds corroding glass, and depositing silica on a wet glass rod (88, 89). *Oxalic Acid*; inflammable gas (carbonic oxide). See column 3.	*b.* Odor of *Hydrocyanic Acid* indicating a *cyanide* or a *ferrocyanide* (blue color) (98, 99, 113).	
b. Brown vapors indicate *Nitric Acid*. See column 3.	*c.* Odor of burning sulphur indicates *Sulphurous Acid* (100) or *Hyposulphurous Acid* (deposition of sulphur (100).	
c. Yellow explosive gas indicates *Chloric Acid* (90).	*d.* No satisfactory result. See col. 3.	
d. Odor of chlorine (91).		
e. Violet vapors indicate *Iodine* (92, 93).	N.B.—If no effect has been observed in cols. 1 and 2, look especially for *Sulphuric* (Table II., col. 1), *Phosphoric* (Table II., cols. 2 and 4), *Boracic* (Table II., col. 6), and *Silicic* (Table II., col. 11) acids.	
f. No definite result. See col. 2.		

* If much heat be applied, thick suffocating fumes of sulphuric acid will be produced, and frequently a smell of sulphurous acid rising from its decomposition.
† Insoluble substances must be examined according to Table I.

TABLE H.

83. Examination of Solutions containing one Acid or Non-metal (not being Organic) (85).

1. Add	2. Add	3. Add	4. Add	5. Add	6. Add
NITRATE OF BARYTA,* White Precipitate insoluble in DIL. NITRIC ACID indicates *Sulphuric Acid* (101, 102).	NITRATE OF SILVER,† White Precipitate insoluble in boiling NITRIC ACID indicates *Hydrochloric Acid* (107, 104, 105).	SULPHATE OF IRON, and pour in very gently STRONG SULPHURIC ACID.	CHLORIDE OF CALCIUM. White Precipitate insoluble in ACETIC ACID, but soluble in HYDROCHLORIC ACID indicates *Oxalic Acid* (114, 115).	one drop of PERCHLORIDE OF IRON. White Precipitate insoluble in ACETIC ACID indicates *Phosphoric Acid* (111, 112).	DIL. HYDROCHLORIC ACID till slightly acid; effervescence indicates *Carbonic Acid* (94, 95);
Yellow Precipitate soluble in DIL. NITRIC ACID indicates *Chromic Acid* (120).	Yellow Precipitate soluble in NITRIC ACID indicates *Phosphoric Acid* (111a, 112).	Brown or Pink ring indicates *Nitric Acid‡* (107, 108, 109).	White Precipitate soluble in ACETIC ACID, add *one drop* of PERCHLORIDE OF IRON. Precipitate indicates *Phosphoric Acid* (111, 112).	Blue Precipitate indicates a *Ferrocyanide* (113).	dip TURMERIC PAPER and dry at a gentle heat. Red color, changed to green by POTASH indicates *Boracic Acid* (116, 117).
White Precipitate soluble in DIL. NITRIC ACID. See 102a.	Yellow Precipitate insoluble in NITRIC ACID and in AMMONIA indicates *Hydriodic Acid* (106). Other Precipitates. See 106a.	Blue precipitate caused by sulphate of iron, insoluble in hydrochloric acid, indicates Ferrocyanogen (113) or Ferrid-cyanogen (p. 72).		Dark Red Color indicates a *Sulphocyanide* (99).	If no result see col. 7.

TABLE II.

	8.	9.	10.	11.
ilk of HLORIC s yellow, ine, and s paper, og cid us Acid 8).	Add SULPHATE OF MANGANESE. Dark brown precipitate indicates *Hypochlorous Acid* (p. 56).	Add SULPHATE OF IRON and POTASH in excess. Shake, and add DIL. HYDROCHLORIC ACID. Blue precipitate indicates *Hydrocyanic Acid.* (99).	Add AMMONIA till it blues red litmus paper, and ACETIC ACID till it reddens blue litmus paper; add very carefully, PERCHLORIDE OF IRON. White, or nearly white precipitate indicates *Phosphoric Acid.* (110), (111), (112).	Add DIL. HYDROCHLORIC ACID till it reddens blue litmus paper, and evaporate to dryness (84). Heat the residue with DIL. HYDROCHLORIC ACID and pour into a test-tube. Semi-transparent flakes indicate *Silicic Acid* (119).

* Chloride of barium may be used, unless silver, lead, or mercury (in the mercurous form) is present.

† Of course nitrate of silver must not be added to a solution which has been made with hydrochloric acid.

‡ It is of course useless to apply this test to a solution which has been made with nitric acid.

NOTES TO TABLES **G** AND **H**.

84. *To evaporate a solution.*—Pour the solution into an *evaporating dish* (*a*, fig. 30), supported on the ring of a *retort-stand* (*b*), and applying a moderate heat.

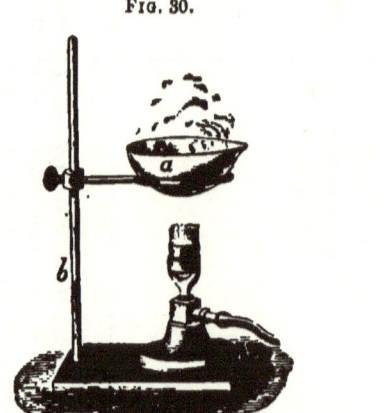

Fig. 30.

Evaporation.

Fig. 31.

If the residue spurts about as the evaporation draws to a close, place the dish upon an empty metal pot (fig. 31) to equalize the heat over its under surface, and let it remain there till thoroughly dry.

Gas-lamps are by far the most convenient for evaporating solutions.

The *Argand Burner* (fig. 32) is one of the best for general

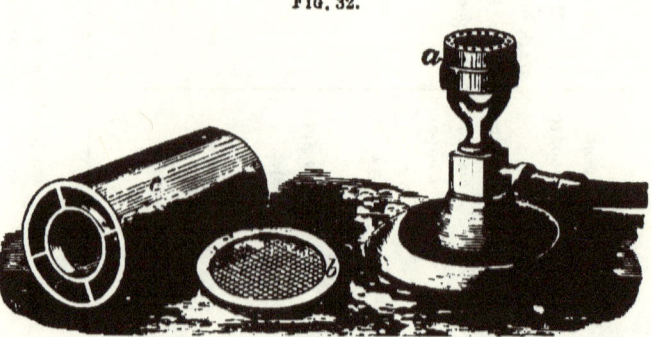

Fig. 32.

use. It should be made so that by unscrewing the burner, a plain jet (fig. 33) for blowpipe experiments may be ob-

Fig. 33.

tained. A brass chimney (*c*, fig. 32) may be made to drop loosely over the burner, resting upon its shoulder (*a*), so as to increase the temperature for some operations. It is also convenient to have a brass ring (*b*), holding a piece of iron wire gauze (with about 400 meshes to the square inch) which may be dropped over the chimney (fig. 34), and the gas lighted above it, so as to obtain the very hot smokeless flame of the mixture of gas and air. It is not advisable, however, to use this flame for small evaporations, since it overheats the sides of the dish and cracks it.

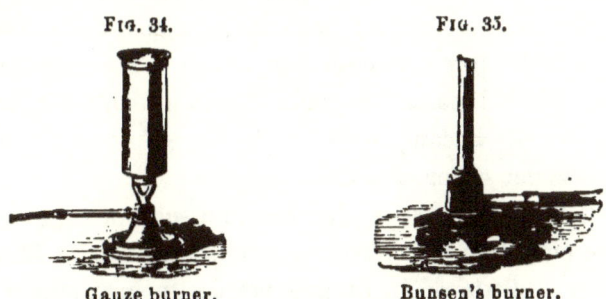

Fig. 34.   Fig. 35.

Gauze burner.   Bunsen's burner.

The *Bunsen's burner* (fig. 35) furnishes a very hot smokeless flame, produced by the admixture of air with the gas. A burner on the same principle may be extemporized with a glass funnel and a plain gas-burner, as described at (70).

By twisting a band of folded paper round the lower part

of the Bunsen's burner, so as to close the air-holes, a luminous flame fit for blowpipe work may be obtained. Such an arrangement also allows the burner to be used with a smaller supply of gas.

Fig. 36.

Where gas is not to be obtained,* a spirit lamp (fig. 36) may be used for evaporation, or even a common candle-flame, if the dish be supported at some little distance above the flame so that it may not be smoked.

85. An organic substance is a substance of animal or vegetable origin; such substances commonly carbonize when heated (17).

86. Since the action of sulphuric acid upon some substances is very violent, care is requisite in this experiment; the test-tube should not be held near the face, and a small quantity of the substance should be employed.

87. Hydrochloric and hydrofluoric acid gases are perfectly transparent in the test-tube, but as soon as they escape into the air, they attract particles of moisture, in company with which they condense into clouds. Hydrobromic and hydriodic acid gases also yield clouds in moist air, but they are generally accompanied by the brown vapor of bromine or the violet vapor of iodine.

88. The presence of fluorine in a substance (giving rise to the evolution of hydrofluoric acid) may be confirmed by placing a little of it, in fine powder, upon a slip of glass, moistening it with strong sulphuric acid, and warming it gently for a minute or two. After cooling, the slip of glass is thoroughly washed, wiped dry, and held so that the eye may glance over its polished surface, when the spot previously

---

* A very convenient supply of portable gas is now furnished by Mr. Orchard, of High Street, Kensington, compressed in safe

occupied by the substance will be found to have entirely lost its polish if fluorine be present.

The action of hydrofluoric acid upon the silica contained in glass, results in the formation of fluoride of silicon gas, which is decomposed when brought into contact with water, depositing opaque silica.

89. If fluorine be detected, the particular form in which it is present remains to be decided.

Uncombined Fluorine, F, if known at all, exists only in the state of gas.

Hydrofluoric Acid, HF, never occurs in the solid state. The commercial acid is a solution in water, always known by its pungent odor and corrosive action on glass.

The only compounds of fluorine which are at all common, are—

Fluor-spar, composed of fluorine and calcium.

Kryolite, which contains fluorine, aluminium, and sodium.

Fluor Spar or *Fluoride of Calcium* has been described at (59).

Kryolite, Na_3AlF_6, is a white opaque mineral, generally in rectangular masses. It is insoluble in water, and but slightly attacked by hydrochloric or nitric acid. The powdered mineral, moistened with hydrochloric acid and exposed on a clean platinum wire (74), colors the flame intensely yellow.

When heated with strong sulphuric acid, kryolite is dissolved; if the solution be further heated, it becomes milky, but the milkiness disappears, if the acid liquid, after cooling, is mixed with much water and boiled.

90. Solid chlorates become yellow or red when moistened with concentrated sulphuric acid, and slowly evolve, even in the cold, a yellow gas (chloric peroxide) resulting from the decomposition of the chloric acid, and having a very peculiar odor.

Great care is necessary in applying heat, since the explosive decomposition of the chloric peroxide sometimes shat-

A solution containing a chlorate becomes yellow when heated with strong hydrochloric acid, emitting an odor resembling chlorine.

Solution of a chlorate will (if free from chloride) give no precipitate with nitrate of silver until it has been acidulated with dilute sulphuric acid and allowed to remain in contact with metallic zinc for a minute or two.

Chloric Acid, $HClO_3$, is not commonly met with. It is a strongly acid liquid, which bleaches test-papers, and evolves an odor of chlorine, especially when heated.

The only chlorates which are at all common, are those of potash and baryta, which have been described at p. 82 and p. 70 respectively.

91. The odor of chlorine is perceived if the substance under examination is common saltpetre, which always contains some chloride. See *Nitre* (p. 81) and Nitrate of Soda (p. 88).

The compounds of hypochlorous acid also evolve chlorine when heated with sulphuric acid.

Hypochlorous Acid itself exists either as a yellow explosive gas, Cl_2O, or an aqueous solution, $HClO(?)$, of strong chlorous odor and great bleaching power.

Solutions of the hypochlorites give a dark brown precipitate of binoxide of manganese on adding sulphate of manganese.

The only hypochlorite commonly met with is the *Hypochlorite of Lime*, $Ca2Clo$, which occurs in the solution of *Chloride of Lime* of commerce, described at p. 74.

Hypochlorite of Soda, $NaClO$, exists in the *Chloride of Soda*, which is sold only in solution, and is described at p. 89.

92. To be sure of the presence of iodine, add to a solution of the substance to be tested a few drops of thin starch (353) and a little concentrated nitric acid; the nitrous acid present in this will liberate the iodine, which colors the starch blue.

93. *Uncombined Iodine*, I, is met with in shining black

chlorine. It stains the fingers brown, and is converted into a splendid violet vapor when heated.

The only *iodides* likely to be met with in ordinary analysis are those of *potassium* (77), *iron* (45), *lead* (14), *mercury* (31), and *arsenic* (35).

94. To recognize the carbonic acid gas, dip a glass rod into lime-water, and introduce it, with the clear drop of lime-water suspended from it, into the mouth of the test-tube, when the drop will immediately become coated with an opaque film of carbonate of lime, which will disappear again if exposed for some time to the action of the gas.

95. *Uncombined Carbonic Acid* is met with in ordinary analysis, either in the state of gas, CO_2, or dissolved in water, $H_2CO_3(?)$, and may be recognized by its faint odor, its feebly reddening blue litmus paper, and its causing a milky precipitate with lime-water, which disappears when the acid is added in excess.

The *carbonates* most commonly met with are those of *lime* (69), *soda* (80), *potash* (77), *ammonia* (75), *baryta* (65), *magnesia* (9), *iron* (45), *zinc* (53), *lead* (14), and *copper* (27).

96. To be sure of the presence of hydrosulphuric acid, spot a piece of filter-paper with solution of acetate of lead or nitrate of silver, which will be blackened when exposed to the action of that gas.

97. *Hydrosulphuric Acid*, H_2S, itself is met with in analysis, either in the form of gas or of a solution in water, which smells of the gas and blackens acetate of lead and nitrate of silver.

Many of the *sulphides* evolve hydrosulphuric acid when they are heated with hydrochloric acid. The most important are those of *iron* (45), *antimony* (39), *ammonium* (75), *lead* (14), *zinc* (53), *potassium*, and *calcium*.

Sulphide of Potassium, K_2S, is generally in brown fragments, which easily become moist on exposure to air, and

water, and the solution generally becomes milky when mixed with hydrochloric acid, from the deposition of a little sulphur, due to the presence either of bisulphide of potassium or of hyposulphite of potash.

Sulphide of Calcium, CaS, occurs in the *soda-waste* of the alkali-works, as a nearly black substance, partly dissolved by boiling water, yielding an alkaline solution. Dilute hydrochloric acid does not entirely dissolve it, but leaves a dark residue containing carbonaceous particles.

Ultramarine is a blue powder which becomes white and evolves hydrosulphuric acid when heated with hydrochloric acid. It contains alumina, silica, sulphur, sodium, and iron.

98. To acquire familiarity with the odor of hydrocyanic (prussic) acid, heat a little solution of ferrocyanide of potassium (yellow prussiate of potash) with dilute sulphuric acid, when pure hydrocyanic acid will be evolved.

In order to be sure of the presence of hydrocyanic acid, add to a solution of the substance a few drops of solution of sulphate of iron and a slight excess of potash; shake the precipitate for a few moments with the air in the tube,* and add an excess of hydrochloric acid, when a blue precipitate, or a decided blue or green color, pervading the liquid, will indicate the presence of hydrocyanic acid, or of a cyanide.

In this test, the ferrous oxide and the potash, acting upon the hydrocyanic acid, produce ferrocyanide of potassium; when the hydrochloric acid is added, it dissolves the ferric oxide produced by the action of the air, and the ferric chloride so produced, coming into contact with the ferrocyanide of potassium, produces ferrocyanide of iron or Prussian blue.

If the cyanogen were originally present as a ferrocyanide or a ferridcyanide, the sulphate of iron would at once produce,

---

* A drop or two of perchloride of iron (ferric chloride) will answer the same purpose as shaking with air.

in the former case, a comparatively light blue precipitate, in the latter, a dark blue.

99. *Hydrocyanic Acid* itself, HCN, is met with in aqueous solution only, recognizable by its odor, its very faintly reddening blue litmus paper, and by the above test.

The principal compounds of cyanogen, which evolve the odor of hydrocyanic acid when they are heated with hydrochloric acid, are *cyanide of potassium* (77), *ferrocyanide* (77), and *ferridcyanide* (77) *of potassium, sulphocyanide of potassium, cyanide of mercury* (31), *fulminate of mercury*.

Sulphocyanide of Potassium, KCNS, forms white needle-like crystals, which become moist in the air and dissolve very easily in water. Perchloride of iron added to the solution gives a deep blood-red color. Heated with hydrochloric acid, the sulphocyanide deposits a yellow precipitate, and evolves a peculiar offensive gas which burns with a blue flame.

Fulminate of Mercury, $HgC_2N_2O_2$, is met with as a grayish crystalline powder which is easily exploded by friction or percussion. If placed on a slip of glass and touched with a lighted match, it burns rapidly with a bright flash, and coats the glass with metallic mercury. It is sparingly dissolved by boiling water, but is readily soluble in hydrochloric acid.

Cyanide of Mercury has been already noticed at (31).

Since potash does not decompose the cyanide of mercury, it is necessary, before applying the Prussian blue test described above, to separate the mercury by slightly acidulating the solution with hydrochloric acid, and introducing a piece of zinc; in the course of a few minutes the solution may be poured off and tested with sulphate of iron, potash, and hydrochloric acid, as above directed.

100. If there be any doubt whether the odor is that of sulphurous acid, place a piece of zinc in the acid liquid, when the hydrogen which is disengaged by its action upon the hydrochloric acid will convert the sulphurous into hydro-

its blackening paper spotted with solution of acetate of lead or nitrate of silver.

Sulphurous Acid itself is met with either as a gas, SO_2, or a solution in water, $H_2SO_3(?)$, always recognizable by its odor. The principal commercial salt of sulphurous acid, the *sulphite of soda*, has been described at p. 89.

Hyposulphurous Acid is not known to exist in an uncombined state. Its only common form of combination, the *hyposulphite of soda*, has been described at p. 89.

101. Small quantities of sulphuric acid (in the form of sulphates) are very commonly found as an impurity in commercial salts, so that if this precipitate be scanty, the analyst must hesitate before pronouncing sulphuric acid to be an essential constituent of the salt. Sulphuric acid will also be detected in a solution which has been made with nitric acid, whether the sulphur existed in the original substance as a sulphide, sulphite, hyposulphite, or sulphate.

102. *Sulphuric Acid* itself (sulphuric anhydride, SO_3), is not commonly met with, except in combination with water.

Hydrated Sulphuric Acid or *Oil of Vitriol*, H_2SO_4, is a heavy, oily liquid which has usually a brownish color due to the presence of organic matter. If it be poured into a little water in a test-tube, much heat is developed.

The *Nordhausen* or *Saxon Sulphuric Acid* emits fumes when the bottle is opened, and hisses slightly when poured into water.

Diluted Sulphuric Acid strongly reddens blue litmus paper. If a piece of white paper be moistened with it, and dried at a gentle heat, it assumes an intensely black color, the paper being carbonized by the acid.

The numerous salts of sulphuric acid or *sulphates* have been described in the notes referring to their respective metals.

102a. A white precipitate by nitrate of baryta, soluble in dilute nitric acid, indicates either carbonic acid (when the precipitate would effervesce with the acid), phosphoric, oxalic, boracic, silicic, sulphurous, or hyposulphurous acid. These will all be detected in the subsequent part of Table II.

103. If possible, the precipitate produced by nitrate of silver should be allowed to settle, and the liquid should be poured off before boiling the precipitate with nitric acid.

Cold nitric acid dissolves all the common precipitates produced by nitrate of silver, except the chloride, sulphide, and cyanide, the two latter requiring to be boiled with the acid.

Cyanide of silver, when washed (16), is dissolved by heating with solution of potash, which converts chloride of silver into the brown oxide of silver, but does not dissolve it.

Several other less common silver precipitates, however, are also insoluble in nitric acid, such as iodide (yellow), bromide, ferrocyanide, ferridcyanide (brown red), and sulphocyanide of silver.

If the washed silver precipitate be shaken with ammonia, the iodide (whitened by the ammonia) and ferrocyanide are undissolved, whilst the others are dissolved by ammonia.

104. Since chlorides are commonly found as impurities in commercial salts, and small quantities give a comparatively large precipitate with nitrate of silver, great care is requisite before concluding that the substance under examination is really a chloride.

The presence of a chloride may be confirmed by heating the original substance with dilute sulphuric acid and black oxide of manganese, when chlorine gas will be evolved, which may be recognized by its odor and by its bleaching moistened litmus paper.*

Bromides would evolve *brown vapors* of bromine having

---

* Corrosive sublimate (mercuric chloride) evolves very little chlorine when heated with diluted sulphuric acid and black oxide of manganese, and does not evolve hydrochloric acid when heated with strong sulphuric acid.

an intolerable odor and imparting an orange color to moist starch.

Iodides would give *violet vapors* of iodine turning moist starch blue.

105. Uncombined *Hydrochloric Acid*, HCl, is usually met with in a state of solution in water. *Concentrated hydrochloric acid*, if pure, is colorless, but the common acid has a yellow color due to iron. It fumes strongly in damp air, and has a peculiar suffocating odor. When heated with black oxide of manganese, it evolves abundance of chlorine, distinguished by its irritating odor and its powerful bleaching effect upon moist litmus paper. *Diluted hydrochloric acid* does not fume in air, but also evolves chlorine when heated with black oxide of manganese.

The *Chlorides*, which are precipitated by nitrate of silver just as hydrochloric acid would be, have been noticed under their respective metals.

106. For a test to confirm the presence of hydriodic acid, see (92).

Hydriodic Acid, HI, in the free state is not commonly met with. The *Iodides*, which behave with nitrate of silver just like hydriodic acid, have been described, when of sufficient importance, under their respective metals.

106a. A *black* precipitate produced by nitrate of silver indicates hydrosulphuric acid (97). A *white* precipitate rapidly changing to *orange brown* and *black* indicates hyposulphurous acid. See *hyposulphite of soda* (80). A *brown* precipitate soluble in nitric acid may be arseniate of silver, indicating arsenic acid (35), or oxide of silver, indicating the presence of some caustic alkaline substance, such as potash, soda, or lime, dissolved in water (69, 77, 80).

A *brown* or *red-brown* precipitate insoluble in nitric acid indicates ferridcyanogen. See ferridcyanide of potassium (77).

A *white* precipitate becoming *brown* or *black* when heated, indicates boracic acid (Table II, col. 6) or sulphurous acid

NITRIC ACID.

A *white* precipitate becoming brown when heated with nitric acid, and then dissolving in ammonia which failed to dissolve it at first, is ferrocyanide of silver (113). If the nitrate of silver be not in excess, a blue color may be produced by the nitric acid. A *red* precipitate indicates chromic acid (120), if the liquid be yellow or red, or arsenic acid if the liquid be colorless. In both cases the precipitate is soluble in nitric acid.

107. This test must be applied to a cold solution; a considerable quantity of sulphate of iron is necessary, and the sulphuric acid must be poured slowly in, so that the bulk of it may sink to the bottom of the tube, for if much heat be produced by its mixing with the water, the brown compound indicative of nitric acid will be decomposed. This brown compound contains sulphate of iron, in combination with nitric oxide which has been formed by the abstraction of oxygen from the nitric acid, in order to convert another part of the sulphate of iron (ferrous sulphate) into the persulphate (ferric sulphate).

108. If additional evidence of the presence of nitric acid be required, the original substance, or even the solution, when cold, may be mixed with about an equal volume of concentrated sulphuric acid, a few copper filings or clippings added, and heat applied, when brown fumes of nitric peroxide will be produced by the deoxidizing effect of the copper upon the nitric acid liberated by the sulphuric acid.*

109. *Nitric Acid* itself (nitric anhydride, N_2O_5) is extremely uncommon, except in combination with water.

Concentrated Nitric Acid, HNO_3, when perfectly pure, is colorless, but it generally has a yellow color caused by the presence of nitric peroxide, NO_2. It fumes in air, stains the

---

* The nitrites, or salts of nitrous acid, evolve brown vapors when treated with sulphuric acid in the cold, and give a brown solution with sulphate of iron and *diluted* sulphuric acid.

skin yellow, and when poured upon copper or zinc causes violent effervescence and disengagement of red fumes.

The *Nitrous Acid* of commerce is concentrated nitric acid containing a larger proportion of nitric peroxide which imparts to it an orange-red color. Some specimens of strong nitric acid have a green color, caused by the presence of nitrous acid.

Diluted Nitric Acid (Aqua Fortis) of course reddens litmus very strongly, and acts upon copper or zinc in the same manner as the concentrated acid, but with less violence.

The salts formed by nitric acid, or *nitrates*, have been described, when important, in the notes relating to the individual metals.

110. If the original solution be neutral (18), it is not necessary to add ammonia and acetic acid, but if it be acid to test-paper, the free acid may prevent the formation of a precipitate with perchloride of iron or with chloride of calcium, so that it must be neutralized with ammonia, which may be added till the liquid smells slightly of it, even after being shaken with the thumb on the top of the tube, and turns red litmus paper blue; acetic acid may be added till the smell of ammonia is no longer perceptible after shaking, and the liquid reddens blue litmus paper. The solution is then examined as in the Table.

Should any precipitate remain undissolved after acetic acid has been added in excess, it is probably phosphate of iron (p. 109), oxalate of lime (58), fluoride of calcium (59), phosphate of alumina (p. 109), or phosphate of lead (p. 110).

111. A single drop of perchloride of iron is here recommended, because the precipitate of phosphate of iron (ferric phosphate) is soluble in an excess of the perchloride.

If a solid compound containing phosphoric acid is throughly dried by heat, and strongly heated in a tube (17) with a little metallic magnesium, the mass, after cooling, evolves the peculiar fishy odor of phosphoretted hydrogen on boiling with water.

The most delicate test for phosphoric acid is *molybdate of ammonia*, which produces, in the solution acidulated with nitric acid (and free from hydrochloric acid), especially on heating, a yellow precipitate containing phosphoric and molybdic acids, and ammonia.

Care must be taken to avoid mistaking arsenic acid for phosphoric acid (p. 46).

111*a*. A solution containing arsenious acid combined with an alkali would also give a yellow precipitate soluble in nitric acid, on addition of nitrate of silver, but the absence of arsenious acid has been previously established (Table A).

112. *Phosphoric Acid* itself (phosphoric anhydride, P_2O_5) is not commonly met with, because it cannot be exposed to air without absorbing moisture and liquefying to a solution of phosphoric acid. The ordinary solution of phosphoric acid is a colorless liquid which strongly reddens blue litmus paper. If it be mixed with a slight excess of ammonia, and with a solution of sulphate of magnesia to which chloride of ammonium and ammonia have been added, a white granular precipitate of phosphate of magnesia and ammonia is produced, the formation of which is much promoted by stirring the liquid (6).

Glacial Phosphoric Acid (or *metaphosphoric acid*, HPO_3) forms transparent colorless masses which easily absorb water from the air. It dissolves in cold water, and the solution gives a white precipitate with nitrate of silver. If the solution of the acid in water be boiled for some time, it is no longer precipitated by nitrate of silver, unless a *very little* ammonia is added, when it gives a yellow precipitate, the metaphosphoric acid having been converted into orthophosphoric or tribasic phosphoric acid, H_3PO_4, by boiling with water.

Phosphoric acid might be found in a solution prepared with nitric acid, as a result of the oxidation of phosphorus.

Ordinary (*vitreous*) *phosphorus* is easily recognized by its inflaming when rubbed or gently heated, when it burns with a bright flame emitting thick clouds of anhydrous phosphoric acid.

Amorphous or red phosphorus does not inflame when rubbed, and requires a higher temperature to inflame it than ordinary phosphorus. When heated in a small tube closed at one end (17) it is converted into vapor which condenses into drops of ordinary phosphorus on the cool sides of the tube.

Phosphate of Soda has been described at (p. 89).

The presence of phosphoric acid in the phosphate of soda may be confirmed by mixing the solution with ammonia, and testing with the mixture of sulphate of magnesia, chloride of ammonium, and ammonia.

The solution of phosphate of soda gives, with nitrate of baryta, a white precipitate which is dissolved by dilute nitric acid.

With nitrate of silver, it gives a yellow precipitate, soluble in nitric acid.

Arseniate of soda, which much resembles the phosphate (p. 46), gives a brown precipitate with nitrate of silver.

Phosphate of Soda and Ammonia, $NaNH_4HPO_4$, or *microcosmic salt*, or *phosphorus salt*, forms colorless crystals, which dissolve in water, yielding a solution which blues red litmus.

With nitrate of baryta and nitrate of silver, the solution behaves like phosphate of soda.

With sulphate of magnesia, especially on stirring, it yields the precipitate of phosphate of magnesia and ammonia.

When microcosmic salt is heated in a dry tube (17), it fuses easily, boils, evolves much steam and ammonia (detected by its odor), leaving a transparent glass of metaphosphate of soda.

Phosphate of Lime and *Superphosphate of Lime* have been noticed at (56).

To confirm the indication of the presence of phosphoric acid in these salts, their solutions may be mixed with acetate of ammonia (prepared by mixing ammonia with a slight excess of acetic acid), and oxalate of ammonia added as long as the precipitate of oxalate of lime is increased. The solution is then boiled, filtered, mixed with ammonia in slight excess, and tested with the mixture of sulphate of magnesia with chloride of ammonium and ammonia.

If phosphate of lime be made into a paste with concentrated sulphuric acid, and exposed on a platinum wire, in the margin of a flame (70), the greenish flame of phosphorus will be perceived in the dark.

Phosphate of Magnesia and Ammonia, or *triple phosphate*, $MgNH_4PO_4$, is a white crystalline powder, which is insoluble in water, but dissolves easily in hydrochloric acid. If the hydrochloric solution be largely diluted with water, mixed with ammonia in excess, and stirred with a glass rod, the phosphate is reprecipitated in a granular form, especially upon the lines of friction made by the glass rod.

The phosphate of magnesia and ammonia evolves ammonia when boiled with potash.

Phosphate of Alumina, $AlPO_4$, is insoluble in water, but dissolves in hydrochloric acid. Potash precipitates it from this solution, but an excess of potash redissolves it. Ammonia also precipitates, but does not redissolve it.

If the solution of phosphate of alumina in hydrochloric acid be mixed with a solution of tartaric acid, and afterwards with an excess of ammonia, the solution will remain clear, and the phosphoric acid may be detected by the mixture of sulphate of magnesia with chloride of ammonium and ammonia.

With strong sulphuric acid, in a flame, phosphate of alumina behaves like phosphate of lime.

Phosphate of Iron, $FePO_4$, is insoluble in water, but soluble in hydrochloric acid, giving a yellow solution. On mixing this with acetate of ammonia (prepared by mixing ammonia with a slight excess of acetic acid), the phosphate of iron is separated as a white precipitate.

If the hydrochloric solution of the phosphate of iron be mixed with ammonia in excess, a brownish precipitate of basic phosphate of iron is obtained; by boiling this with sulphide of ammonium, the iron is converted into (black) sulphide of iron, and the phosphoric acid is dissolved as phosphate of ammonia, and may be detected in the filtered solution by sulphate of magnesia, mixed with chloride of ammonium and ammonia.

Phosphate of Lead is white, insoluble in water, and not easily soluble in hydrochloric acid, but diluted nitric acid dissolves it; ammonia added to this solution precipitates the phosphate of lead, and acetic acid does not redissolve it.

By boiling the phosphate of lead with sulphide of ammonium, the phosphoric acid is converted into phosphate of ammonia, and may be detected in the filtered solution, with the mixture of sulphate of magnesia with chloride of ammonium and ammonia.

113. *Ferrocyanogen* is the name given to the group C_6N_6Fe containing carbon, nitrogen, and iron, which is supposed to exist in the ferrocyanide of potassium, and in other ferrocyanides, though it has never been obtained in the separate state.

Hydroferrocyanic (or *ferrocyanic*) *Acid*, $H_4C_6N_6Fe$, is not at all a common substance. It is crystalline, and easily soluble in water. When the solution is boiled, it evolves the odor of hydrocyanic acid, and deposits a white precipitate of cyanide of iron, which rapidly turns blue by oxidation.

Ferrocyanide of Potassium has been described at (p. 84). Its solution gives, with nitrate of silver, a nearly white precipitate of ferrocyanide of silver, which is not dissolved by ammonia.

If this precipitate be warmed with dilute nitric acid, it is converted into the orange-red ferridcyanide of silver, and is then dissolved by ammonia, with exception of a few flakes

Sulphate of iron causes a blue precipitate in solution of ferrocyanide of potassium.

Ferrocyanide of Iron, or *Prussian Blue*, $Fe_5 12CN$, is not dissolved by water or dilute acids. It may be identified by boiling it with potash, which leaves brown peroxide of iron undissolved, and yields a solution of ferrocyanide of potassium which may be filtered, neutralized with hydrochloric acid, and tested with perchloride of iron.

Ferrocyanide of Copper, or *Hatchett's Brown*, $Cu_2 Fe6CN$, has a brown-red color, and is insoluble in water, and in dilute acids. It is decomposed by boiling with potash, into black oxide of copper and ferrocyanide of potassium, which may be tested as in the case of Prussian blue.

114.* In order to obtain confirmatory evidence of the presence of oxalic acid, shake a little of the original substance with diluted sulphuric acid (which does not cause effervescence with oxalic acid), and throw in a little powdered binoxide of manganese, when carbolic acid will escape with effervescence, especially if a gentle heat be applied, and may be recognized by the lime-water test (94).

115. *Oxalic Acid*, $H_2C_2O_4.2Aq.$, itself forms colorless prismatic crystals, which dissolve easily in water, and yield a strongly acid solution.

When heated with strong sulphuric acid, oxalic acid dissolves with effervescence, evolving carbonic acid and carbonic oxide gases, the latter of which burns with a blue flame on approaching the mouth of the tube to a flame.

With nitrate of baryta, a strong solution of oxalic acid gives, on stirring with a glass rod, a granular precipitate of oxalate of baryta, which dissolves in diluted nitric acid.

With nitrate of silver, solution of oxalic acid gives a white

---

* Ferrocyanide of potassium may also give, with chloride of calcium, a precipitate of the double ferrocyanide of calcium and potassium, which might be mistaken for oxalate of lime, since it is insoluble in acetic acid, but soluble in hydrochloric acid, from which ammonia reprecipitates it.

precipitate of oxalate of silver, which is soluble in diluted nitric acid.

When heated in a dry tube (17), oxalic acid melts very easily, and is entirely converted into vapor, a part of which condenses on the sides of the tube in fine transparent needles.

The Oxalates commonly met with are the *acid oxalate of potash* (p. 85), *oxalate of ammonia* (75), and *oxalate of lime* (58).

116. This test for boracic acid depends upon its singular property of coloring orange-red the dye known as turmeric; but this property belongs only to free boracic acid, so that if it be combined with a base, it is necessary to liberate it by adding hydrochloric acid, until the solution reddens blue litmus paper. A large excess of hydrochloric acid must be avoided, since it is liable to carbonize the paper on drying.

The paper may be dried by waving it above the flame so as not to scorch it, or by gently warming it upon a slip of glass.

To obtain confirmatory evidence of the presence of boracic acid, mix the substance under examination with strong sulphuric acid; add alcohol, and inflame the mixture, either upon a glass rod dipped into it, or on a slip of glass, or in a small evaporating dish, in the latter case stirring it with a glass rod. Boracic acid imparts a decided green color to the flame, especially at the edges.

117. *Boracic Acid* itself (*anhydrous boracic acid*, B_2O_3) is not commonly met with. It forms glassy fragments, transparent when freshly prepared, but becoming opaque when kept. It is powdered with great difficulty, and appears at first to be insoluble in water and acids, but after boiling with water for a short time, a little of it is dissolved, yielding a solution which turns blue litmus paper violet, and stains turmeric paper, which has been dipped into it and dried, orange-red, becoming green when moistened with potash.

Glassy boracic acid dissolves easily when heated with potash.

Crystallized (or *hydrated*) *Boracic Acid*, $3H_2O.B_2O_3$, forms white scaly or feathery crystals, which dissolve in boiling water, and are deposited again when the solution cools. The solution behaves, with litmus and turmeric papers, as stated above. Placed on the point of a knife or on a platinum wire (74), and heated in the margin of a flame (70), crystals of boracic acid color the flame green, a part of the acid being converted into vapor in the presence of the water. The crystals are dissolved by boiling alcohol, and the solution burns with a fine green flame.

The impure boracic acid imported from Tuscany, always contains considerable quantities of ammonia and sulphuric acid, as impurities.

The only common borates are *Borax*, or *Biborate of Soda* (p. 89), and the mineral Boronatrocalcite, which is composed of boracic acid, soda, lime, and water, and is imported from Peru under the name of Borate of Lime.

Boronatrocalcite* occurs in soft earthy masses; it is dissolved to a slight extent by water, and entirely by hydrochloric acid. If it is dissolved in a small quantity of hydrochloric acid, ammonia will cause a precipitate of borate of lime; but if much acid be present, the chloride of ammonium formed will be sufficient to prevent the precipitation, since borate of lime is soluble in chloride of ammonium.

118. To detect silicic acid, the solution which has been acidified with hydrochloric acid is evaporated to *dryness†* (i. e., until no more liquid remains) in a dish (84); the latter is allowed to cool, and some diluted hydrochloric acid poured into it. The dish is again gently heated, and its contents stirred with a glass rod; when nothing more appears to dissolve, the contents of the dish are poured into a test-tube. If silicic acid is present, semi-transparent flakes will be

---

* $Na_2O.2B_2O_3 2(CaO.2B_2O_3).18Aq.$

† When a considerable quantity of silicic acid is present, the liquid becomes a jelly at a certain stage of the evaporation.

seen in the liquid, giving it a peculiar opalescent appearance.

The silicic acid may be filtered off, and the solution tested for potassium and sodium according to Table F, p. 77.

The only common substance which is likely to be mistaken for silicic acid in this test is sulphate of lime, which is not immediately redissolved by hydrochloric acid after evaporation to dryness, though continued heating with the acid will dissolve it. But the sulphate of lime is more opaque, and is generally left as a crystalline powder, which sinks much more readily than the silicic acid. Should there be any doubt, the insoluble part may be allowed to settle in the tube, washed twice, by decantation (16) and heated with a little potash, which will easily dissolve silicic acid, but not sulphate of lime.

119. The common forms of silicic acid, as well as the more frequently occurring silicates, are insoluble in water and in diluted acids, and will therefore be noticed hereafter.

The only silicates which are soluble in water are the *Silicate of Potash* (p. 85), and *Silicate of Soda* (p. 90). Even those silicates which are attacked by boiling with diluted acid (such as the *Zeolitic* minerals, and various slags), do not generally dissolve in the acid, but leave a residue of gelatinous silicic acid, so that they will be noticed together with substances insoluble in water and acids.

120. *Chromic Acid* itself, CrO_3, forms needle-like crystals of a crimson color; they attract moisture readily from air, and dissolve easily in water; the solution gives a bright yellow precipitate with acetate of lead, and a red precipitate with nitrate of silver.

The chromates most commonly met with are *chromate of potash* (61), *bichromate of potash* (61), *chromate of baryta*

EXAMPLES FOR PRACTICE. 115

121. *Examples for Practice in Tables* **G** *and* **H.**—The following substances may be analyzed for practice (10) :—

Chalk (carbonate of lime)	Oxalic acid
Common salt (chloride of sodium)	Saltpetre (nitrate of potash)
Fluor spar (fluoride of calcium)	Chlorate of potash
Chloride of lime	Iodide of potassium
Sulphide of iron	Cyanide of potassium
Ferrocyanide of potassium	Sulphite of soda
Hyposulphite of soda	Sulphate of magnesia
Phosphate of soda	Borax (biborate of soda)

Silicate of potash.

EXERCISE VII. (See (141) for Examples for Practice.)

122. Analysis of Insoluble Substances which may contain one Metal and one Acid or Non-metal (not being Organic) 85, 125a.

TABLE I.

1.	2.	3. Fuse the powdered substance (124, 125), on a piece of platinum foil* with about 3 parts of CARBONATE OF SODA for some minutes (130). Soak the foil with the mass in warm water (131), and filter.			
		Solution.			Residue.†
		1.	2.	3.	Wash with water till the washings no longer give a precipitate with chloride of barium; treat with hot DILUTE NITRIC ACID, and afterwards with water.
Heat on a piece of porcelain (123).	Heat the powdered (124) substance with a little STRONG SULPHURIC ACID.	Acidulate with HYDROCHLORIC ACID, and add CHLORIDE OF BARIUM. Precipitate not dissolved by more Hydrochloric Acid indicates *Sulphuric Acid* (132, 133).	Acidulate with HYDROCHLORIC ACID, and evaporate to dryness (118). Warm the residue with DIL. HYDROCHLORIC ACID, and pour the solution into a test-tube. Insoluble flakes *Silicic Acid* (137).	Acidulate with HYDROCHLORIC ACID and add excess of HYDROSULPHURIC ACID. Yellow Ppt. indicates *Tin* (134, 135). Orange Ppt. indicates *Antimony* (136).	Solution. Examine for a metal according to Table A. (135).
Sulphur melts easily, burns blue, and emits the smell of sulphurous acid (126). *Carbon* does not melt, but glows when strongly heated, and burns away (127). *Chloride of Silver* melts into brown drops, which remain unchanged (128). *Fluor Spar* crackles and emits a transient blue light (129).	Pungent fumes depositing opaque silica upon a wet glass rod indicate a *Fluoride* (86, 89).				Residue. Wash, and treat with hot DIL. HYDROCHLORIC ACID. Solution. Examine for the metal according to Table A. (138). Residue. Gelatinous *Silicic Acid*, or (gritty) undecomposed substance. (139, 140).

* The platinum foil is occasionally corroded in this operation, so that a small piece only is used.
† If this residue is gelatinous, it probably consists of silicate of alumina.

EXPLANATIONS AND INSTRUCTIONS ON TABLE I.

123. *Mode of observing the action of heat upon substances in contact with air.*—A small piece of platinum foil is very convenient for this purpose.

It may be held in a pair of crucible tongs (fig. 37), or upon a triangle of iron wire (fig. 38), one limb of which is thrust through a cork for convenience in holding it when hot.

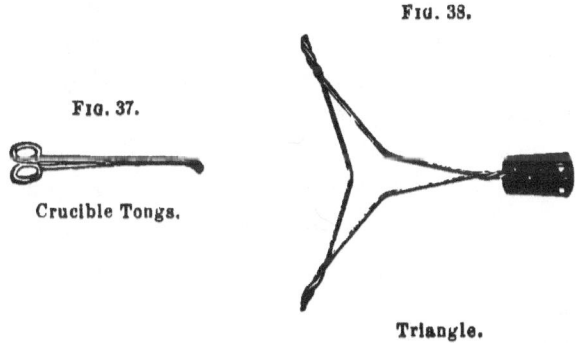

Fig. 37. Crucible Tongs.

Fig. 38. Triangle.

The substance under examination should be placed at one corner of the foil, so that the latter may not be rendered entirely useless if the substance should corrode it.

A gentle heat should be applied at first by holding the foil a little above the point of a flame. A stronger heat may be afterwards obtained by directing the blowpipe-flame upon the under surface of the foil (fig. 39).

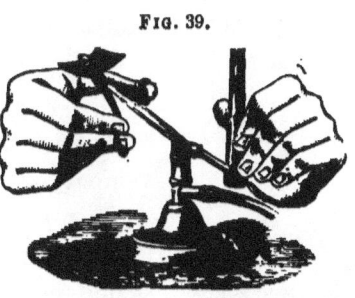

Fig. 39.

Compounds of lead, bismuth, tin, antimony, and arsenic should not be heated on platinum foil, since these substances corrode it.

A piece of broken porcelain, or a small porcelain crucible, may be used instead of platinum foil.

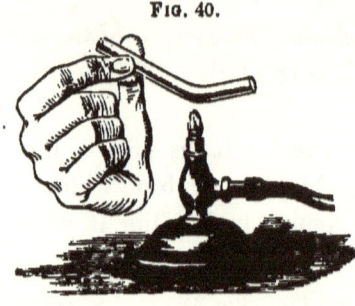

Fig. 40.

A small piece of thin German glass tube open at both ends will answer the same purpose, especially if it be slightly bent so that the substance may rest in it. The tube should be so held that one opening is considerably higher than the other, and to this higher opening the nose should be applied, in order to detect the odor of sulphur, arsenic, &c. (fig. 40).

124. *Reduction of substances to powder.*—Most substances can be crushed and afterwards ground to powder in a pestle and mortar (fig. 41) of Wedgwood ware. A small spatula (fig. 42) is used to scrape the powder out of the mortar.

Fig. 41.

Fig. 42.

Spatula.

Large fragments must first be broken up by the hammer on an anvil.

Fig. 43.

An iron pestle and mortar are very convenient for crushing hard substances.

To prevent fragments of the substance from flying about, a wooden cover (fig. 43) should be placed over it, with a hole for the handle of the pestle.

When the mortar is not at hand, substances of moderate hardness may be powdered on a hard surface by rolling a thick bottle over them.

An agate mortar and pestle (fig. 44) are used for the final grinding of very hard substances. The agate pestle should be provided with a handle made of hard wood, and fitted to it by a brass cap (fig. 45). The powdering is very much

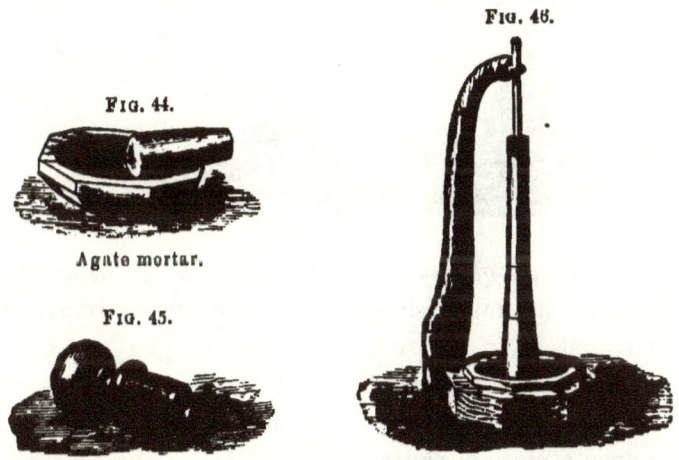

Fig. 44.

Agate mortar.

Fig. 45.

Fig. 46.

facilitated by mounting the pestle on a brass rod, which plays in a ring attached to a stout wooden upright rising from a heavy bed of hard wood in which the agate mortar is firmly set (fig. 46).

Reduction of substances to powder is sometimes hastened if the fragments are sifted out from time to time, by rubbing the powder lightly with the finger upon a piece of muslin tightly stretched over the mouth of a beaker (fig. 47). The fragments left on the muslin are returned to the mortar.

Fig. 47.

For powdering small quantities of very hard substances, a

steel diamond mortar (fig. 48) is employed, consisting of a socket (*a* fig. 49), into which a steel cylinder open at both ends (*b*) fits tightly. The substance to be powdered, in small fragments, is dropped into the hollow cylinder, and the solid steel cylinder (*c*) which fits the former exactly, being placed upon it, it is struck sharply and repeatedly with a heavy hammer.

Fig. 48.   Fig. 49.

125. *Fusion of insoluble substances with carbonate of soda.*—The substance should be reduced to an impalpable powder, that is, to a powder in which no grit can be felt (124), and intimately mixed with the carbonate of soda, also in fine powder, either in a small mortar (fig. 41) or with a knife upon a piece of paper.

The carbonate of soda employed for this fusion must have been previously dried to expel its water of crystallization. The fusion is easier when a mixture of carbonates of soda and potash is employed.

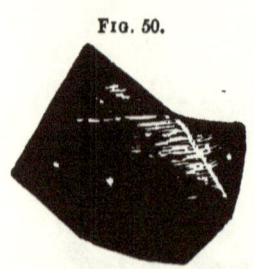

Fig. 50.

The mixture is placed upon a piece of platinum foil,* slightly bent up at the edges (fig. 50), and heated moderately at first to expel any moisture which may be present. The foil is then heated to redness, over a Bunsen (fig. 51) or gauze (fig. 52) burner, or by directing a broad blowpipe-flame upon

* A piece of foil $1\frac{1}{2}$ in. long and $1\frac{1}{4}$ in. wide, of such thickness as to weigh seven grains, will be found suitable for the purpose.

its under surface (fig. 39).* To obtain such a flame, the blowpipe jet should be held a little away from the flame, so that a broad divergent stream of air may be sent through it.

Fig. 51.

Gauze burner.

Fig. 52.

Bunsen's burner.

Since only a small quantity of the powder can be fused at once upon the foil, it is advisable to add it in successive portions, as each is melted, until a sufficient quantity of the substance has been employed.

When the mixture is very difficult to fuse, the blowpipe flame must be directed down upon its surface after the lower part has been fused. No violence must be employed to detach the fused mass from the foil, but it should be soaked in water as directed at (131).

A small platinum capsule (fig. 53) is very convenient for the fusion of insoluble substances.

Fig. 53.

NOTES TO TABLE I.

125*a*. A substance insoluble in water and acids will *most likely* be one of the following:—carbon, sulphur, silica, silicate of alumina (clay), fluoride of calcium, sulphate of

---

* It is better to support the platinum foil upon the triangle (fig. 38) than to hold it with the tongs, which generally contaminate the substance with iron.

baryta, sulphate of lime, binoxide of tin, chloride of silver, sulphate of lead or chrome-iron-ore.

126. Sulphur is likely to be met with in several forms.

Crude Sulphur, as imported from Sicily and elsewhere, forms grayish yellow lumps, and leaves a slight dark residue when burnt.

Distilled Sulphur, used for the manufacture of gunpowder, forms pure yellow lumps, and burns without residue.

Roll Sulphur (*common brimstone*) forms pure yellow cylindrical sticks, also burning without residue.

These three varieties dissolve entirely, or nearly so, in bisulphide of carbon.

Flowers of Sulphur (or *sublimed sulphur*) is a fine yellow powder, which burns without residue, but is not entirely dissolved by the bisulphide of carbon, since it contains a considerable proportion of the insoluble variety of sulphur.

Milk of Sulphur (or *precipitated sulphur*) is a white impalpable powder which generally leaves a considerable white residue (sulphate of lime) when burnt.

Viscous Sulphur, obtained by pouring hot melted sulphur into water, has a brown color, and somewhat resembles India-rubber; it becomes brittle when kept for a few hours. Sulphur sometimes separates in viscous masses when the sulphides of the metals are dissolved in nitric acid.

127. Carbon is met with in the forms of diamond, graphite, vegetable charcoal, animal charcoal, coal, coke, gas-carbon, lamp-black, soot.

Diamond is generally recognized by its extreme hardness, rendering it capable of scratching glass and even steel. It is unaffected when heated, in a small leaden or platinum cup, with hydrofluoric acid, or a mixture of sulphuric acid and ammonium fluoride, which would dissolve the imitations of diamond. It burns with difficulty in air, even when heated in the blowpipe-flame, and is best recognized by burning it in oxygen, and detecting the carbonic acid produced, by means of lime-water. The mode of effecting the combustion of the diamond is described in most works on Elementary Chemistry.

Graphite, *plumbago*, or *black lead* is recognized by its semi-metallic lustre, especially when rubbed with a hard substance, and by its greasy feeling between the fingers. It burns away very slowly when heated in air, and generally leaves a considerable residue of (brown) peroxide of iron.

Vegetable Charcoal (wood charcoal) may of course generally be recognized by its appearance. It glows readily when heated in air, and leaves a small quantity of white very light ash.

Animal Charcoal (*bone-black*, *ivory-black*, *char*) burns pretty easily when strongly heated, but leaves a very large earthy residue (phosphate and carbonate of lime). This variety of charcoal effervesces slightly with hydrochloric acid, which dissolves the phosphate and carbonate of lime, the former being deposited as a white gelatinous precipitate on mixing the filtered solution with ammonia in excess.

Coal evolves a tarry odor when heated, and furnishes a coke which burns away with some difficulty, leaving a moderate ash consisting chiefly of silica and alumina, but having a reddish-brown color (due to peroxide of iron) if much iron-pyrites be present in the coal.

Coke of course behaves in the same way, but does not evolve tarry odors. Both coal and coke burn vividly (deflagrate) when thrown into melted nitre heated to redness.

Gas-carbon (the deposit which lines the interior of gas-retorts) somewhat resembles coke in appearance, but is much harder and more compact, having almost a metallic sound when struck. It is very difficult to burn it in air, and a high temperature is required to deflagrate it with nitre.

Lamp-black (*spirit-black*) is known by its dead black appearance and great lightness. It glows easily when heated in air, and burns entirely away when pure, but commercial lamp-black often leaves a considerable white ash.

Soot may be recognized by its odor. It burns easily in air. When boiled with solution of potash, it evolves an odor of ammonia, derived from the destructive distillation of the coal.

128. *Chloride of Silver*, AgCl, dissolves when heated with ammonia; if an excess of nitric acid be added to the solution, white chloride of silver is reprecipitated, and becomes violet when exposed to daylight.

Chloride of silver also dissolves in solution of hyposulphite of soda.

129. The presence of fluor-spar may be confirmed according to (59).

130. Should the fused mass have a bright yellow color, it is due to the formation of an alkaline chromate, and indicates the presence of chromium. A dark green color, changing to turquoise blue on cooling, is caused by manganate of soda, and indicates the presence of manganese.

131. The platinum foil with the fused mass should be placed, as soon as the mass has set, in an evaporating dish (fig. 54) containing about half an ounce (four teaspoonfuls) of distilled water. The dish is gently heated, and the surface of the foil lightly rubbed under the water, with a glass rod rounded at the end, until the mass has become detached. The foil is then removed, and the mass stirred with the water, and crushed under the glass rod until it is entirely disintegrated; the insoluble residue is then collected on a filter.

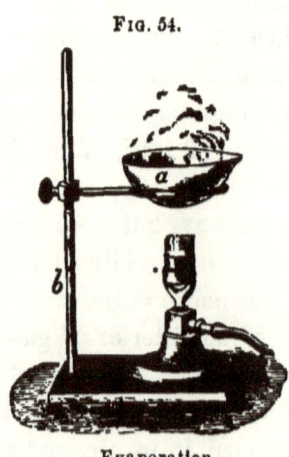

Fig. 54.

Evaporation.

132. Since the carbonate of soda (which has been used in the fusion) is liable to contain sulphuric acid as an impurity, it is advisable to test it for that acid before concluding that the sulphuric acid here detected really belongs to the substance.

If the sulphuric acid be really present as an essential constituent of the substance insoluble in water and acids, it must exist in the form of sulphate of baryta, sulphate of strontia, sulphate of lime, or sulphate of lead.

133. *Sulphate of Baryta*, $BaSO_4$, or *heavy-spar* (sometimes called *barytes* or *cawk*) occurs naturally in a transparent colorless form, crystallized in prisms. It is more commonly opaque, and of a brownish color. By a skilful operator the barium may be detected in it by the blowpipe test (276).

Artificial sulphate of baryta is an earthy white powder.

Sulphate of Strontia, $SrSO_4$, or *celestine* usually occurs in bluish opaque masses with a fibrous structure, but it is sometimes white or yellowish. The blowpipe test will indicate the strontium (244).

Sulphate of Lime and *Sulphate of Lead* have been described at (69) and (14) respectively.

134. It sometimes happens that a slight brown precipitate of platinum sulphide is produced here, due to the corrosion of the platinum foil during the fusion.

135. If tin be present in the insoluble substance, it will probably also be detected in the residue left on treating the fused mass with water.

The only compound of tin likely to be found here is the *Binoxide of Tin*, SnO_2 (*tin stone, stannic acid, putty-powder*), which occurs naturally either in separate crystals (stream-tin ore) or in large masses (mine-tin ore) of inferior purity. It is usually dark-colored, heavy, and very hard, scratching glass like quartz.

The presence of tin is easily confirmed by the blowpipe (241).

Artificial binoxide of tin is a white powder which becomes yellowish when heated.

Putty-powder commonly contains oxide of lead as well as binoxide of tin.

136. If antimony is present in the insoluble substance, the greater part of it will probably be found, as antimoniate of soda, in that portion of the fused mass which is insoluble in water.

Antimonic Acid, Sb_2O_5, the only insoluble compound of antimony likely to be met with here, is a white powder which assumes a yellow tint when heated.

137. For confirmatory evidence of the presence of silicic acid, see (118, 119).

Silicon or silicium itself, Si, would have been converted into silicic acid by the fusion with carbonate of soda. Its commonest form is that of a dark powder which dissolves when boiled with potash. If a little silicon be placed on a piece of platinum foil and sharply heated by directing a blow-pipe-flame upon the under surface of the foil, it eats a hole in the metal, converting it into the fusible silicide of platinum.

The so-called *oxide of silicon* which occurs in the residue left when cast iron is dissolved in hydrochloric acid, is a gray, very light powder, which is easily dissolved with effervescence when heated with solution of potash, hydrogen being evolved, which is recognized by its inflammability.

Silicic Acid or silica, SiO_2, is met with in a variety of forms, of which the following are the most important.

Sand is of various shades of color. It is generally found to contain a small quantity of alumina (in the form of clay) and a little iron.

Flint is known by its characteristic appearance.

Quartz occurs in rounded pebbles or in transparent masses, sometimes of a pink color. It is also commonly met with in well-defined six-sided prisms with pyramidal terminations. Both quartz and flint scratch glass easily.

Chalcedony varies very much in color, its commonest form is milk-white and opaque.

Soluble Silica is found in dull white earthy masses, or, in volcanic districts, in porous lumps like pumice, often stained yellow with perchloride of iron. It is easily dissolved when boiled with solution of potash.

COMMON SIMPLE SILICATES.

Names.	Composition.
Clay	Silicic acid, alumina, water.
Pumice	Silicic acid, alumina.
Slate	Silicic acid, alumina.
Steatite	Silicic acid, magnesia.
Meerschaum	Silicic acid, magnesia.
Iron slags	Silicic acid, oxide of iron.
Electric calamine	Silicic acid, oxide of zinc, water.

Clay* (silicate of alumina) occurs in various degrees of purity, and may be generally recognized by its plasticity when mixed with water.

By heating clay for some time with strong sulphuric acid, a part of it is decomposed, and if, after cooling, the mixture be diluted with water and filtered, it gives, when mixed with excess of ammonia, the characteristic gelatinous precipitate of alumina.

Pipe-clay and *Kaolin* are white, and consist of nearly pure silicate of alumina.

Fire clay (Stourbridge clay) has a gray color, and contains a little iron. *Fire-brick* (baked fire-clay) has a yellowish hue, from the presence of ferric oxide.

Dinas fire-brick consists almost entirely of silica.

Common Clay has various shades of blue, yellow, and red, and often contains considerable quantities of iron and lime.

Fuller's Earth is a brown clay containing iron.

Brick, earthenware, and porcelain, since they are composed chiefly of baked clay, contain silicate of alumina, the two former sometimes containing much iron.

When clay is fused, as directed in the table, the greater

---

* Kaolinite, which appears to form the basis of the varieties of clay, has the composition $Al_2O_3.2SiO_2.2Aq$.

part of the silica is not found in the aqueous solution of the fused mass (as is the case with most other silicates), but in a gelatinous residue of silicate of alumina which is left undissolved by water, but dissolves in acids.

Pumice and *Slate* are known by their appearance.

Steatite, $3MgO.4SiO_2$, *Soap-stone*, or *French Chalk* (silicate of magnesia), is recognized by its peculiar soapy feel when rubbed in the fingers.

Meerschaum, $2MgO.3SiO_2.2Aq.$, is a white earthy mineral, met with in rounded masses.

Iron Slag (silicate of iron) has been described at (p. 56).

Electric Calamine (hydrated silicate of zinc), or *zinc glance*, $2ZnO.SiO_2Aq.$, is usually grayish-white, with a glassy lustre. When the mineral is boiled with hydrochloric acid, the silica separates in the gelatinous state, and the zinc is dissolved in the form of chloride.

138. It is of course unnecessary to examine for potassium, sodium, and ammonium in this solution.

139. Any portion of the substance which has not been finally powdered is likely to be left here.

140. Chrome-iron ore is never completely attacked by the fusion, and may be recognized by fusing it on platinum foil with carbonate of soda and nitrate of potash. On soaking the foil in warm water, a yellow solution is obtained which should be filtered, acidified with acetic acid, and tested with lead acetate, which gives a yellow precipitate. The undissolved residue may be dissolved in strong hydrochloric acid and tested for iron (44).

Carbon in some very incombustible forms, may also be left here. Such carbon can only be consumed by protracted heating in a muffle.

141. *Examples for Practice in Table* **I.**—The following substances may be analyzed for practice. See (10).

Fluor spar	Sulphate of baryta
Plaster of Paris	Sulphate of lead
Binoxide of tin	White sand.

EXERCISE VIII. (See (162) for Examples for Practice.)

142. Analysis of Substances which may contain one Metal and one Organic Acid.*

TABLE K.

1.	2.	3.	4.
Heat a little of the substance on a piece of broken porcelain, or in a tube open at both ends (17). Blackening, and evolution of some peculiar smell (146) indicates the presence of organic matter. Continue the heat until all the carbon is consumed. If no residue is left see (147). Moisten the residue, and test with RED LITMUS PAPER. If this be colored blue, *Potassium,* *Sodium,* or *Calcium* is probably present. See column 2.	Heat a little of the substance in a test tube with CONCENTRATED SULPHURIC ACID. Immediate blackening indicates the probable presence of *Tartaric Acid.* See Table L. Gradual blackening, with evolution of carbonic oxide (burning with a blue flame) indicates the probable presence of *Citric Acid.* See Table L. Evolution of much carbonic oxide (burning with a blue flame) without blackening, indicates the probable presence of *Oxalic Acid* (115), a *Cyanide* (99), or a *Ferrocyanide* (113), or *Formic Acid* (150).	Examine a portion of the substance for the metal by Table A.	If the substance is soluble in water, examine the solution for the acid according to Table L. If it is insoluble in water, proceed as in Table N.

* Excluding those acids which are comparatively rare, such as racemic, those which are recognized by some conspicuous physical property, such as the fatty acids, butyric acid, etc., and those for which no direct test can be given, such as lactic acid.

TABLE L.

143. Examination for an Organic Acid in an Aqueous Solution.

Test the solution with BLUE and RED LITMUS PAPERS.

A. *The Solution is Neutral.*—Examine by columns 2, 3, 5, and 6.
B. *The Solution is Acid.*—Examine by columns 1, 2, 3, 4, 5, 6.
C. *The Solution is Alkaline.*—Examine by Table M.

1.	2.	3.	4.	5.	6.
To a part of the Solution add a little POTASH, so as to leave it still acid, and stir briskly with a glass rod (73). Crystalline precipitate indicates *Tartaric Acid*. See column 3.	To another part of the solution add a little PERCHLORIDE OF IRON.* Red color indicates *Acetic Acid* (148, 149), or *Formic Acid* (150), or *Meconic Acid* (151), or *Hydrosulphocyanic Acid* (99). Inky Black Color or precipitate indicates *Tannic Acid* (152), or *Gallic Acid* (153). Brown precipitate indicates *Benzoic Acid* (134), *Succinic Acid* (155), or *Hippuric Acid* (156).	To a fresh part add AMMONIA in slight excess, filter, if necessary, and add CHLORIDE OF CALCIUM, and stir. Precipitate, add ACETIC ACID. Dissolved *Tartaric Acid*, (157, 158), undissolved, *Oxalic Acid* (114, 115). If no precipitate, boil the solution: precipitate indicates probable presence of *Citric Acid* (159).	To a fresh part add CARBONATE OF AMMONIA *very carefully* till the solution is *neutral* to test papers, filter, if necessary, and add a little PERCHLORIDE OF IRON. See column 2. Mistakes are common here, because the carbonate of ammonia precipitates the perchloride of iron, if it be added in excess (159*a*).	To a fresh part add ACETATE OF LEAD, white precipitate becoming crystalline on standing, and fusing so as to stick to the bottom and sides of the tube when the liquid is boiled, indicates *Malic Acid* (160). (Tartaric, oxalic, citric, and meconic acids are also precipitated by lead acetate.)	To a fresh part add NITRATE OF SILVER, and boil. Precipitate of metallic silver indicates probable presence of *Formic Acid* (150). If there is no metallic precipitate, add ammonia drop by drop to the Boiling solution. Be sure there should be an excess of acid, which would prevent the precipitation of silver. Test for hydrocyanic acid, or a cyanide, as at (98), (161*a*).

* A blue precipitate indicates a ferrocyanide (113); a dark brown color indicates a ferridcyanide (77).

144. Examination for an Organic Acid in an Alkaline Solution.

1.	2.	3.
To a part of the solution add DIL. HYDROCHLORIC ACID in slight excess,* and stir briskly with a glass rod. White pulverulent precipitate, probably *Uric Acid* (161). Feathery crystalline precipitate, probably either *Benzoic Acid* or *Hippuric Acid* (154, 156).	To another part add ACETIC ACID in *slight* excess, and ACETATE OF POTASH.† Stir briskly. Crystalline precipitate, presence of *Tartaric Acid* (158).	To another part add DIL. NITRIC ACID till the solution is very slightly acid, and examine by columns 3, 4, 5, and 6 of Table L.

* A *very* slight excess of hydrochloric acid may cause a precipitate of bitartrate of potash, which is easily redissolved by further addition of hydrochloric acid.
† Made by adding acetic acid to a solution of potash till it slightly reddens blue litmus paper.

TABLE N.

145. Examination for an Organic Acid in a Substance insoluble in Water.

1.	2.
Boil a portion of the substance with CARBONATE OF SODA, filter if necessary, and examine for the acid by Table M.	Place a little of the substance in a dish, or on a piece of platinum foil, or a slip of glass, add some DILUTE NITRIC ACID, and evaporate to dryness. Yellow residue, becoming red when treated with ammonia, and purple with potash, indicates *Uric Acid* (161).

NOTES TO TABLES K, L, M, AND N.

146. Much attention should be paid to the odor evolved on heating the substance, since many organic acids may be at once recognized in this way. The analyst is recommended to render himself familiar with the characteristic odors by heating small specimens of each of the organic acids.

Tartaric Acid evolves a sweetish odor, somewhat like that of *burnt sugar*.

Citric Acid evolves a similar odor, but more pungent.

The compounds of *Acetic Acid* give the peculiar fragrant odor of acetone.

Benzoic Acid gives the aroma of *frankincense*.

Succinic Acid emits vapors which provoke *coughing*.

Hippuric Acid furnishes an odor resembling *bitter almonds*.

Uric Acid evolves a smell of *singed hair*, in which a keen scent will trace ammonia and hydrocyanic acid.

147. If the substance leaves no residue when heated, it is not likely to contain any metal except ammonium or mercury.

Ammonium may at once be sought by boiling the substance with solution of potash, which would evolve the pungent odor of ammonia.*

Mercury would be detected by boiling the substance with dilute hydrochloric acid and slips of metallic copper, which would become coated with a silvery deposit, yielding globules of mercury when dried and heated in a small tube.

Proceed to detect the acid, as in column 2, Table K.

148. The red color caused by acetic acid (due to the formation of peracetate of iron or ferric acetate) should not

---

* It must be remembered that urea also evolves ammonia when boiled with potash; but urea, unlike ammonia, is not precipitated

be bleached by chloride of mercury (in which it differs from that caused by the hydrosulphocyanic acid).

To obtain confirmatory evidence of the presence of acetic acid, heat the substance with alcohol and concentrated sulphuric acid, which should evolve the pleasant smell of acetic ether, resembling that of cider.

149. *Acetic Acid*, $HC_2H_3O_2$, itself is a colorless liquid, having the acid *smell of vinegar*, without its aroma. It evaporates when heated on a slip of glass, without leaving any residue.

If acetic acid is carefully neutralized with ammonia or carbonate of ammonia, and a little nitrate of silver added, a crystalline precipitate of acetate of silver separates, especially on stirring briskly with a glass rod. The silver is not reduced to the metallic state on heating the precipitate with the liquid (as would be the case with formic acid, which, in other respects, might be mistaken for acetic).

Acetic acid yields a crystalline precipitate with mercurous nitrate (protonitrate of mercury), if the mixture be stirred briskly with a glass rod.

Acetic acid is also characterized by its property of acquiring an alkaline reaction to red litmus, when boiled with oxide of lead (litharge), in consequence of the formation of the basic acetate of lead.

The most commonly occurring salts of acetic acid are Acetate of Lead (14), Tribasic Acetate of Lead, Acetate of Copper (27), Acetate of Alumina (51), Acetate of Iron, Acetate of Potash, Acetate of Soda, and Acetate of Ammonia.

Tribasic Acetate of Lead (or *Goulard's extract*), $Pb2C_2H_3O_2.2PbO.Aq.$, is commonly sold in solution, though it may be crystallized in needles. It has a sweet taste, and turns red litmus paper blue. On breathing into a test-tube containing a little of the solution, and shaking it up, a white precipitate of carbonate of lead is produced. Addition of common water to the solution also renders it milky.

Acetate of Iron (ferric acetate), $Fe3C_2H_3O_2$, forms a deep red solution, which is changed to yellow by hydrochloric acid. When a solution of ferric acetate is mixed with three or four times its volume of water and boiled, a brown precipitate of basic ferric acetate is deposited.

Acetate of Potash, $KC_2H_3O_2$, is commonly sold in white opaque fibrous masses, having a faint acetic odor, and easily absorbing moisture from the air. It is very soluble in water, and the solution gives a bright red color with perchloride of iron.

When heated on platinum foil, acetate of potash fuses easily, evolves inflammable vapors, and leaves a residue of carbonate of potash, which is strongly alkaline to moistened red litmus paper, and effervesces with hydrochloric acid.

Acetate of Soda, $NaC_2H_3O_2.3Aq.$, is sold either in moist transparent crystals or in grayish fibrous masses which have been fused. In other respects it resembles acetate of potash, but the residue of carbonate of soda, which it leaves when burnt on platinum foil, does not deliquesce in air like the carbonate of potash. It is very fusible, and requires a high temperature to decompose it.

Acetate of Ammonia, $NH_4C_2H_3O_2$, is generally met with in solution; on boiling it, the odors of acetic acid and ammonia may be perceived.

150. *Formic Acid*, $HCHO_2$, is a colorless liquid of pungent odor, resembling that of acetic acid. It leaves no residue when evaporated. When formic acid is heated with nitrate of silver or bichloride of platinum, black deposits of the metals are obtained, especially if a little ammonia is added.

The formiates of potassium and sodium, when heated with strong sulphuric acid, evolve carbonic oxide abundantly, which burns with a blue flame on applying the mouth of the tube to a light.

151. *Meconic Acid,* $H_3C_7HO_7.3Aq.$, forms scaly crystals which have commonly a brownish color, and dissolve with difficulty in cold, but easily in hot water. Alcohol also dissolves them.

The red color produced by perchloride of iron with meconic acid is not bleached by chloride of mercury (thus distinguishing it from hydrosulphocyanic acid).

A solution of meconic acid which has been mixed with excess of ammonia, gives a white precipitate with chloride of calcium, which is soluble in acetic acid.

None of the meconates are of common occurrence. Meconate of morphia is sometimes used in medicine.

152. *Tannic Acid* (or tannin), $C_{14}H_{10}O_9$, is a yellowish or brownish powder, which has a very astringent taste, and dissolves pretty easily when shaken with cold water.

On adding potash or ammonia to a solution of tannic acid, it becomes brown, especially if shaken with air.

Dilute sulphuric acid, added in considerable quantity, causes a white precipitate in solution of tannic acid. When the solution has been kept for some time, it fails to give this precipitate.

The only tannate commonly met with is that of sesquioxide of iron (ferric tannate) which exists in writing ink.

Ink may be identified by adding strong nitric acid, which reddens it, and changes it to a clear yellow on boiling. On adding excess of ammonia, the rusty brown sesquioxide of iron is precipitated.

153. *Gallic Acid,* $HC_7H_3O_5$, is a white or yellowish powder composed of minute shining needles. It is not visibly dissolved when shaken with cold water, but dissolves readily on boiling, being deposited again in delicate needles when the solution cools.

Potash or ammonia causes a red-brown color in a solution of gallic acid, which becomes much darker on shaking with air.

There is no gallate of common occurrence.

154. Be quite sure that this brown precipitate is not the peroxide of iron precipitated because the original solution was alkaline.

Benzoic Acid, $HC_7H_5O_2$, is commonly met with in white *feathery crystals*, which are extremely light and have an agreeable odor of incense. It has a peculiar sweetish pungent taste, and is not easily dissolved by water. It dissolves easily in ammonia, and is reprecipitated in feathery flakes on adding excess of hydrochloric acid, if the solution be not too dilute.

When gently heated in a tube, it fuses easily, and sublimes without leaving any residue of importance. Hippuric acid (156) much resembles benzoic.

None of the benzoates are sufficiently common to require special description.

155. The precipitate is of a much darker red-brown color in the case of succinic than in that of benzoic acid.

Succinic Acid, $H_2C_4H_4O_4$, is a colorless crystalline acid, which dissolves easily even in cold water. The solution when mixed with excess of ammonia and chloride of barium, does not give any precipitate; but if alcohol be added, the liquid deposits succinate of baryta as a white precipitate.

Succinic acid is characterized by its behavior when heated; it first melts, and is then decomposed, evolving vapors which produce violent coughing. It leaves a very slight carbonaceous residue, which easily burns away.

No salt of succinic acid can be described as being commonly met with.

156. *Hippuric Acid*, $HC_9H_8NO_3$, forms shining prismatic crystals which are nearly insoluble in cold water, but dissolve readily on boiling, and are deposited as the liquid cools. It is almost insoluble in ether, and may thus be distinguished from benzoic acid, which dissolves easily in ether.

When heated in a tube, hippuric acid melts and is decom-

resembling oil of bitter almonds, and leaving a carbonaceous residue.

When heated with strong sulphuric acid, hippuric acid blackens more easily, and evolves more sulphurous acid than benzoic acid does. If hippuric acid be dissolved in potash and evaporate to dryness, the residue evolves ammonia when heated. Benzoic acid, being free from nitrogen, could not furnish ammonia when thus treated.

The hippurates are not commonly met with.

157. The behavior of the precipitate of tartrate of lime is very characteristic, and should be carefully observed.

When precipitated, as in Table L, from an ammoniacal solution, it is flocculent or gelatinous, according to the strength of the solution; but if it be set aside for some time, it becomes a granular crystalline precipitate.

The flocculent tartrate of lime dissolves easily in chloride of ammonium, and is deposited again in a crystalline state if the sides of the tube beneath the liquid be well rubbed with a glass rod.*

Crystalline tartrate of lime does not dissolve when shaken with chloride of ammonium, and not easily in acetic acid, so that it might at first be mistaken for oxalate of lime; but if it be dissolved in hydrochloric acid, and an excess of ammonia added, it is not immediately reprecipitated, as would be the case with oxalate of lime, but requires brisk stirring with a glass rod, when it again separates in a crystalline state.

158. *Tartaric Acid*, $H_2C_4H_4O_6$, itself is sold either in colorless crystals, or as a white powder. It dissolves readily, even in cold water, giving a strongly acid solution.

When heated, tartaric acid melts, carbonizes, and evolves a peculiar odor, somewhat resembling that of burnt sugar;

---

* The tartrate of lime precipitated from tartar emetic does not exhibit this disposition to become crystalline.

the remaining carbon burns off without any residue if the acid is pure.

When heated with strong sulphuric acid tartaric acid soon gives a very black solution, and evolves a little carbonic oxide, which burns with a blue flame on applying the mouth of the tube to a light.

Lime-water, added in excess to a very small quantity of solution of tartaric acid, gives a white precipitate of tartrate of lime, soluble in chloride of ammonium.

A solution of tartaric acid, mixed with a *small quantity* of chloride of calcium, gives no precipitate; but if potash be gradually added, tartrate of lime is precipitated, which dissolves in an excess of potash, and is reprecipitated on boiling the solution, dissolving again as it cools.

If much chloride of calcium be employed, some carbonate of lime is precipitated by the carbonic acid in the potash, and this does not dissolve in the excess of potash.

Acetate of lead, added to a solution of tartaric acid, produces a white precipitate of tartrate of lead, which is easily soluble in ammonia.

If a solution containing tartaric acid be mixed with potash in excess and permanganate of potash, the green color at first produced will disappear on boiling, a brown precipitate of manganic oxide being separated (see citric acid).

The salts of tartaric acid commonly met with, are Bitartrate of Potash, Tartrate of Antimony and Potash, and Tartrate of Potash and Soda.

Bitartrate of Potash has been described at (p. 83).

By dissolving it in a very small quantity of potash, and adding a little chloride of calcium, the white precipitate of tartrate of lime may be obtained, which dissolves in an excess of potash, and is reprecipitated by boiling.

Tartrate of Antimony and Potash has been noticed at (39).

Its solution in water is slightly acid, and gives, after a short time, a white precipitate of teroxide of antimony on adding ammonia; this may be filtered off, and the solution tested with chloride of calcium (157).

Tartrate of Potash and Soda, or Rochelle salt, $KNaC_4H_4O_6.4Aq.$, forms large transparent prismatic crystals, which dissolve easily in water. If the solution be slightly acidified with acetic acid, and briskly stirred, a crystalline precipitate of bitartrate of potash is deposited.

159. The precipitation by chloride of calcium in an ammoniacal solution on boiling, must not be regarded as satisfactory proof of the presence of citric acid, since other precipitates, particularly carbonate and sulphate of lime, may, under some conditions, be obtained in a similar way.

Citric Acid, $H_3C_6H_5O_7.Aq.$, forms colorless crystals, which are readily dissolved by cold water.

When heated, the crystals fuse, carbonize, and emit irritating inflammable vapors, very different from those evolved by tartaric acid. Citric acid also leaves less charcoal than tartaric. The charcoal burns away entirely if the acid is pure.

Strong sulphuric acid heated with citric acid, evolves much carbonic oxide gas, recognized by its burning with a blue color on applying the mouth of the tube to a flame; the mixture does not carbonize so readily as in the case of tartaric acid.

Lime-water, added in large excess to a very small quantity of solution of citric acid, does not produce a precipitate until the mixture is boiled, when citrate of lime is deposited. On cooling the solution, the precipitate will, at least partly, be redissolved, since citrate of lime is more soluble in cold than in hot water.

gives no precipitate; but if potash be added, citrate of lime is separated, which does not redissolve, like tartrate of lime, on adding an excess of potash.

If a solution containing citric acid be mixed with potash in excess and permanganate of potash, the latter will be changed to the green manganate of potash, which will not become brown on boiling (see tartaric acid).

The only citrates which are at all commonly met with are the medicinal preparations—citrate of iron, ammonio-citrate of iron, citrate of iron and quinine, and citrate of bismuth.

Citrates of lime and magnesia are imported from Sicily, for the preparation of citric acid.

Citrate of Iron (ferric citrate) forms brilliant red or yellow transparent scales, which taste sweet and astringent; they dissolve readily in water, forming a brown solution which is precipitated by alcohol. The presence of citric acid prevents the precipitation of the ferric oxide on addition of ammonia.

Ammonio-citrate of Iron is very similar to the citrate.

Citrate of Iron and Quinine also forms fine yellow scales, in which, however, the bitter taste of quinine is perceptible. Its solution yields a white precipitate of quinine on addition of ammonia.*

Citrate of Bismuth is remarkable as being the only preparation of bismuth which can be dissolved in water without decomposition.

By decomposing its solution with hydrosulphuric acid, the bismuth may be precipitated as sulphide, and the filtered solution, after evaporating to expel excess of hydrosulphuric acid, may be tested for citric acid.

---

* The citrate of iron and quinine of the *Pharmacopœia* contains both ferrous and ferric oxides, so that it yields a dark blue color with both ferrocyanide and ferricyanide of potassium.

In the medicinal preparation, citrate of ammonia is usually present.

Citrate of Lime, $Ca_3 2C_6H_5O_7.4Aq.$, is a white powder which turns red litmus blue. It dissolves sparingly in water, but readily in hydrochloric acid. Ammonia precipitates the solution only when boiled. When heated on platinum, it is charred and decomposed, leaving a residue of carbonate of lime, which effervesces with hydrochloric acid.

Citrate of Magnesia, $Mg_3 2C_6H_5O_7.14Aq.$, is a crystalline granular salt, which is more soluble in water than the citrate of lime.

The so-called "granulated effervescing citrate of magnesia" is often composed of a mixture of carbonate of soda with citric, or occasionally tartaric, acid, which do not act upon each other until water is added.

159a. The chance of error from this cause is diminished by mixing the original solution with carbonate of ammonia until it is slightly alkaline, and evaporating at a gentle heat till it is neutral.

160. This is a very satisfactory test for mallic acid, if the solution be not too dilute and a large excess of lead acetate be avoided. The malate of lead fuses into a white opaque drop, which feels sticky when touched with a glass rod.

The crystallized mallic acid is somewhat deliquescent, has an agreeable sour taste, and dissolves easily in water and alcohol. When heated in a tube it fuses easily, and is afterwards decomposed, with effervescence, evolving very pungent vapors of fumaric and malæic acids, which condense in crystals upon the cool part of the tube.*

161. The presence of uric acid may be confirmed by the test given in column 2 of Table N.

Uric Acid (or *lithic acid*), $H_2C_5H_2N_4O_3$, itself is a white or yellowish white crystalline powder, which dissolves very

---

* The malic acid of commerce sometimes contains a little oxalic

sparingly, even in boiling water. It is also nearly insoluble in hydrochloric acid, but dissolves on boiling in dilute nitric acid, with brisk effervescence. It is also dissolved when heated with a moderately strong solution of potash, and is reprecipitated from this solution in a crystalline state, on adding a slight excess of hydrochloric acid.

When heated, uric acid is carbonized, and emits vapors in which the smell of ammonia and that of hydrocyanic acid can be distinguished.

On platinum foil, uric acid, if pure, burns without residue.

Strong sulphuric acid does not blacken uric acid to any great extent, even on heating, but dissolves it, in great measure without change. On allowing the solution to cool, and adding water, the greater part of the uric acid is precipitated.

The uric acid deposited from urine and occurring in urinary calculi, is often colored yellow or brown by coloring matters derived from the urine.

The urates of soda and ammonia are commonly met with as deposits from urine, colored yellow, brown, or red, by the coloring matters of that excretion. They are both dissolved when heated with water, and deposited again as the solution cools. On adding hydrochloric acid to the warm solution, crystalline uric acid is precipitated.

Urate of Ammonia, $NH_4HC_5H_2N_4O_3$, of course, evolves the smell of ammonia when boiled with potash.

Urate of Soda, $NaHC_5H_2N_4O_3$, when heated on platinum foil, leaves a residue of carbonate of soda, recognized by its solubility in water, giving a strongly alkaline solution, which effervesces with hydrochloric acid.

161a. In some cases it is difficult to identify the

Where the present metal belongs to either of the first three classes (p. 18), the salt may be suspended or dissolved in water, treated with an excess of hydrosulphuric acid, filtered from the metallic sulphide, evaporated to a small bulk, and tested by Table L. Salts of barium, strontium, calcium, and magnesium may be boiled with carbonate of soda, filtered, and the solution examined by Table M.

Some of the rarer organic acids are described at (174, 215, 218, 232).

162. *Examples for Practice in Tables* **K, L, M,** *and* **N.**—The following substances may be analyzed for practice (10):—

Tartaric acid	Tannic acid
Bitartrate of potash	Gallic acid
Tartar emetic	Benzoic acid
Acetate of lead	Citric acid
Uric acid	Acetate of soda.

EXERCISE IX. (See (171) for Examples for Practice.)

163. Identification of the more frequently occurring Vegetable Alkaloids.*

TABLE O.

Dissolve the substance in a *small* quantity of WATER or of DILUTE HYDROCHLORIC ACID.

1.	2.	3.	4.	5.
To a part of the solution add *diluted* POTASH, very carefully, till the solution is very slightly alkaline. Stir with a glass rod, and set aside. If a precipitate is produced, add POTASH in excess. Dissolved. *Morphine* (163). | Undissolved. See Column 2.	To another part of the solution add, if necessary, DILUTE SULPHURIC ACID till slightly acid, and a saturated solution of BICARBONATE OF SODA till the liquid no longer reddens blue litmus. Stir briskly with a glass rod, and set aside. If no precipitate, test first by column 3. If a precipitate is produced, test first by column 4, and afterwards by column 3.	Moisten some of the original solid, in a porcelain dish, with STRONG SULPHURIC ACID. Red color indicates *Brucine* (166). Add a minute quantity of BICHROMATE OF POTASH on the end of a glass rod. Dark purple color indicates *Strychnine* (167).	To another part of the original solution add some *strong* CHLORINE-WATER and AMMONIA in excess. Green color indicates *Quinine* (168). Orange red color indicates *Narcotine* (169).	To another part of the original solution add AMMONIA in excess, and shake with ETHER. Precipitate undissolved by the ether, probably *Cinchonine* (170). Test for Caffeine as in (164).

* Heat a small portion of the substance in a small tube open at both ends.
If a *fine red liquid* is deposited on the sides of the tube, and the *remarkable odor* of quinoline (resembling that of *tar*) is perceived, either quinine or cinchonine is probably present (see Table O, columns 4 and 5).
If a *very powerful odor of ammonia* is perceived, without much blackening, some base of animal origin, such as urea (152), may be suspected.

NOTES TO TABLE O.

164. Caffeine would give no precipitate with potash, and would have escaped detection in the Table.

Caffeine (or Theine), $C_8H_{10}N_4O_2.Aq.$, crystallizes in needles, which have a somewhat bitter taste, and dissolves slightly in cold water, but entirely in boiling water; the solution is neutral. It is also soluble in alcohol and ether. Gently heated in a tube, caffeine fuses and sublimes in fine needles. To identify it, dissolve it in very little strong hydrochloric acid, add a small crystal of chlorate of potash, and evaporate just to dryness: the residue has a pink or red color, and dissolves in ammonia to a fine purple liquid, which is bleached by potash.

165. *Morphine* $C_{17}H_{19}NO_3.Aq.$, is a white crystalline powder, which dissolves sparingly even in boiling water, yielding a bitter solution, which is alkaline to test-papers. It is also soluble in alcohol, but not in ether.

Strong nitric acid colors morphine orange-yellow, and produces a similar color in solutions containing morphine.

Perchloride of iron (ferric chloride), free from excess of acid, colors morphine inky blue.

Morphine, heated with strong sulphuric acid, and stirred with a glass rod moistened with nitric acid, gives a rapid play of colors, from dingy green to a rich brown.

The hydrochlorate (muriate), meconate, and acetate of morphine are easily soluble in water. By mixing the aqueous solution with carbonate of soda, and stirring briskly, the morphine is precipitated, and may be collected upon a filter, washed with cold water, and tested with perchloride of iron or with nitric acid.

Meconate of Morphine is not crystallizable.

Acetate of Morphine, $C_{17}H_{19}NO_3 \cdot C_2H_4O_2$, is crystalline, but very deliquescent.

Hydrochlorate of Morphine is crystalline, $C_{17}H_{19}NO_2 \cdot HCl \cdot 3Aq.$, and not deliquescent.

The acids may be detected as in Tables H and L.

166. *Brucine*, $C_{23}H_{26}N_2O_4 \cdot 4Aq.$, is a white crystalline powder, which may be dissolved by boiling water, yielding a bitter solution. It also dissolves in alcohol, but not in ether.

Strong nitric acid gives a bright red solution with brucine which becomes yellow when heated. Protochloride of tin (stannous chloride) changes the yellow color to violet.

If brucine be dissolved in a drop or two of diluted hydrochloric acid, and ammonia carefully added, an oily-looking precipitate of brucine separates, which afterwards changes to needle-like crystals. Excess of ammonia dissolves the oily precipitate, and the solution deposits the needles after some time.

167. *Strychnine*, $C_{21}H_{22}N_2O_2$, is met with either as a fine white powder, or in hard prismatic crystals. It is scarcely perceptibly dissolved, even by boiling water, but the solution has an intensely bitter taste. It dissolves rather sparingly in ordinary alcohol, and is soluble in absolute alcohol and in ether.

If strychnine be dissolved, on a white surface of porcelain, in strong sulphuric acid, and stirred with a few particles of binoxide of lead (brown oxide), it gives a dark violet purple color, quickly changing to red. The test with bichromate of potash, given in Table O, is far more delicate, especially if a very minute quantity of the bichromate be employed.

168. A solution of bromine in water may be substituted for chlorine water in testing for quinine. Strong solutions of quinine yield a green precipitate when tested with chlorine (or bromine) and ammonia. Another excellent test for quinine consists in adding to the acid solution a little chlorine or bromine water, a drop or two of ferridcyanide of potassium, and, drop by drop, ammonia, which produces a fine red color bleached by excess of ammonia.

Quinine, $C_{20}H_{24}N_2O_2.3Aq.$, is a white crystalline powder, which is sparingly dissolved even by hot water, but dissolves easily in alcohol. Ether does not dissolve it so easily. Its solution is very bitter, and is alkaline to test-papers.

If quinine be dissolved in diluted sulphuric acid, and the solution mixed with water and examined by daylight in a test-tube, it will be found to exhibit a very pretty shade of blue when in certain positions, though it appears quite colorless when held directly between the eye and the light.

This *fluorescence* is very characteristic, and may be seen even in dilute solutions.

Sulphate (or basic sulphate) *of quinine*, $2C_{20}H_{24}N_2O_2.H_2SO_4.7Aq.$, forms very light silky needles, which are very bitter, and will not dissolve, even in boiling water, unless a little sulphuric or hydrochloric acid is added.

If sulphate of quinine is adulterated with salicine, $C_{13}H_{18}O_7$, it assumes a red color when moistened with strong sulphuric acid.

To detect the presence of cinchonine, the sulphate is shaken in a test-tube (or small stoppered bottle) with ammonia and ether, when pure sulphate of quinine entirely dissolves, the solution separating into two layers; whilst any cinchonine, being insoluble in ether, separates on the surface of the lower (aqueous) layer.

169. *Narcotine*, $C_{22}H_{23}NO_7$, is a white crystalline tasteless substance, which is not alkaline to moistened test-papers. It is insoluble in water, but dissolves in alcohol and ether, yielding bitter solutions.

To identify it, dissolve it in a considerable quantity of strong sulphuric acid, and stir the liquid with a glass rod moistened with strong nitric acid, a dark red color is produced, which is bleached by more nitric acid.

If a few drops of solution of perchloride of iron be carefully added, from the end of a glass rod, to the solution of narcotine in strong sulphuric acid, a deep red liquid is produced, which becomes of a brighter color on cooling.

If narcotine be dissolved in dilute hydrochloric acid, and a little solution of bromine be added, a yellow precipitate is obtained, unless the solution is very dilute. On heating, this precipitate is dissolved, and, by gradually adding solution of bromine and boiling, a fine rose color is produced, even in very dilute solutions. Excess of bromine destroys the color.

170. *Cinchonine*, $C_{20}H_{24}N_2O$, is white, crystalline, and slightly bitter. It is almost insoluble in water, but dissolves in alcohol, yielding a bitter solution which has an alkaline reaction. Ether does not dissolve it. Gently heated in a tube, cinchonine fuses, emits a peculiar tarry ammoniacal odor, and yields a sublimate of shining needles on the cooler part of the tube. Moistened with dilute sulphuric acid, and heated, it yields a fine red coloring matter.

If cinchonine be dissolved in as little dilute hydrochloric acid as possible, the solution gives, with ferrocyanide of potassium, a yellow precipitate, which dissolves when warmed with a slight excess of the ferrocyanide, and is deposited in yellow scales or needles on cooling.

Sulphate of cinchonine, $2C_{20}H_{24}N_2O.H_2SO_4.2Aq.$, forms white or brownish prismatic crystals which fuse when heated, and yield a fine red coloring matter, as well as an aromatic odor.

171. *Examples for Practice in Table* **O.**—The following substances may be examined for practice (10):—

Hydrochlorate of morphine	Sulphate of quinine
Strychnine	Cinchonine.

EXERCISE X. (See (234) for Examples for Practice.)

172. IDENTIFICATION OF THE MORE COMMON ORGANIC SUBSTANCES.*

1. SOLID ORGANIC SUBSTANCES.

A. *Characterized by color.*

173. *Indigo*, C_8H_5NO.—Dark blue. Insoluble in water, alcohol, and ether.

Heated in a tube (17), yields violet vapors, smelling of aniline and ammonia.

Strong sulphuric acid slowly dissolves indigo, when heated, giving a blue solution, which is changed to brown-yellow by nitric acid.

Shaken in a corked tube with sulphate of iron (ferrous sulphate) and slaked lime, and allowed to settle, indigo dissolves to a yellow solution (reduced indigo, C_8H_6NO), which becomes blue-green when decanted and acidulated with hydrochloric acid.

174. *Picric* or *Carbazotic Acid*, $HC_6H_2O3NO_2$.—Yellow crystals. Very bitter. Stains the skin yellow.

Water dissolves it sparingly. Bright yellow solution.

Alcohol dissolves it easily. The solution gives a yellow crystalline precipitate when stirred with a little potash.

Heated in a tube, fuses, and sometimes explodes slightly.

Strong sulphuric acid dissolves it, and deposits it unchanged on addition of water.

Heated with solution of chloride of lime (bleaching powder), it evolves a very pungent odor like oil of mustard, due to chloropicrine, CCl_3NO_2.

175. *Caramel*, $C_{12}H_{18}O_9$.—Dark-brown. Deliquescent. Slightly bitter.

Very soluble in water, dark-brown solution.

Sparingly soluble in strong alcohol.

Heated in tube, carbonizes, and emits the odor of burnt sugar.

Strong sulphuric acid carbonizes it.

B. *Characterized by Odor.*

176. *Carbolic Acid (Phenic Acid); Phenole*, C_6H_6O.

Moist needle-like crystals; colorless or pale brown. Powerful odor of coal-tar. Very easily melted. Water dissolves it sparingly. Easily soluble in potash.

Alcohol dissolves it readily.

Perchloride of iron (ferric chloride) gives a dark purple-blue color with the aqueous solution of carbolic acid.

Dropped into strong nitric acid, carbolic acid is oxidized with great violence, yielding a red solution; if this be boiled and allowed to cool, it deposits prismatic crystals of picric acid which may be identified by (174).

176*a.* *Hydrate of Chloral*, $C_2HCl_3O.H_2O.$, white crystalline solid. Easily dissolved by water.

Remarkable pungent odor.

Easily melted and volatilized.

Heated with potash, yields an oily-looking layer of chloroform and a solution of formiate of potash, which may be identified by neutralizing with dilute sulphuric acid, adding nitrate of silver in excess, decanting the clear liquid from any precipitated chloride of silver, and boiling it, adding ammonia, drop by drop, when metallic silver is precipitated.

C. *Without Characteristic Color or Odor.*

Examine by Table P.

TABLE P.

77. Examination of a Solid Organic Substance which cannot be distinguished by its Color or Smell, or by the Tests for Organic Acids and Alkalies given in Tables K, L, M, N. O.

1.	2.	3.	4.
Shake a portion of the substance with cold WATER for some time. If it dissolves, either ntirely or in great measure, examine the solution by (178). If it does not dissolve, pass on to column 2.	Boil the substance with WATER. If it dissolves, examine the solution by (186). If it does not dissolve, pass on to column 3.	Boil the substance with ALCOHOL. If it dissolves, examine by (193). If it does not dissolve, pass on to column 4.	Boil the substance with ETHER (200). If it dissolves, examine by (201). If it does not dissolve examine by (206).

NOTES TO TABLE P.

178. The commonest of these organic substances which are dissolved to any considerable extent by being shaken with cold water are—*

Cane-sugar	Pyrogalline (or pyrogallic acid)
Grape-sugar (or glucose)	Salicine
Milk-sugar (or lactine)	Urea
Soluble albumen.	

179. *Cane-sugar*, $C_{12}H_{22}O_{11}$, identified.

Add to the solution a few drops of sulphate of copper, and, drop by drop, potash. The blue precipitate first produced redissolves in the excess of potash, to a blue liquid. Boil for some minutes, suboxide of copper (cuprous oxide) is deposited, first as a yellow hydrate, afterwards as the red anhydrous oxide.

To another part of the solution, add excess of potash, and boil, only a very light-brown color should be produced.

Heat a portion of the solid substance with strong sulphuric acid, which should carbonize it almost immediately.

To a part of the aqueous solution, add a few drops of dilute hydrochloric acid, and boil for a few minutes; the cane-sugar is thus converted into grape-sugar, which may be identified by the following tests.

180. *Grape-sugar*, $C_6H_{14}O_7$, identified.

Add to the solution an excess of potash, and boil. The liquid should assume a rich brown color.

To another part of the solution, add a few drops of sulphate of copper, and, drop by drop, potash. The blue precipitate first produced redissolves in the excess of potash, to a blue liquid, which deposits suboxide of copper when heated (at first yellow and afterwards red) more readily than in the case of cane-sugar.

Heat a portion of the solid substance with strong sulphuric acid: it should not carbonize so readily as cane-sugar.

Grape-sugar is much less sweet than cane-sugar.*

181. *Milk-sugar*, $C_{12}H_{24}O_{12}$, identified.

Very much less sweet than either cane- or grape-sugar.

Feels gritty between the teeth. Almost insoluble in ordinary alcohol, which dissolves cane- or grape-sugar on heating.

Answers to the same tests as grape-sugar.

182. *Urea*, $C_2H_4N_2O_2$, identified.

Prismatic crystals. Resembles nitre in appearance and taste. Very soluble in water and alcohol.

Heated in a tube, melts easily, and evolves much ammonia.

A pretty strong aqueous solution of urea, stirred with concentrated nitric acid in excess, and allowed to stand, deposits scaly crystals of nitrate of urea.

If a strong solution of oxalic acid be substituted for nitric acid, crystals of oxalate of urea are deposited.

If a solution of urea be mixed with a solution of mercuric nitrate,† a white precipitate is produced.

On adding excess of silver nitrate to solution of urea, and boiling down, in a test-tube, to a small bulk, a crystalline precipitate of silver cyanate separates on cooling; if this precipitate be heated with dilute hydrochloric acid, it effervesces, and evolves the pungent odor of cyanic acid; on adding potash, and boiling, the odor of ammonia is perceived.

183. *Pyrogallic acid* or *Pyrogallin*, $C_6H_6O_3$, identified.

White or slightly brown crystalline powder. Often

---

* Sugar of fruits (fructose or uncrystallizable sugar) answers to the same taste as grape-sugar.

† Prepared by adding finely-powdered red oxide of mercury to hot nitric acid as long as it is dissolved.

collected into feathery flakes. Bitter. Very light. Dissolves easily in water, alcohol, and ether.

Potash renders it intensely brown, oxygen being absorbed from the air.

Sulphate of iron (ferrous sulphate) colors solution of pyrogallic acid dark blue.

Perchloride of iron (ferric chloride) gives a fine red color with solution of pyrogallic acid.

184. *Salicine*, $C_{13}H_{18}O_7$, identified.

White crystalline powder. Bitter. More soluble in alcohol than in water. Insoluble in ether. Concentrated sulphuric acid converts it into a blood-red resinous mass. Its aqueous solution, boiled for some time with hydrochloric acid, yields a granular precipitate of saliretine.

185. *Soluble albumen*, $C_{72}H_{112}N_{18}SO_{22}$, identified.

Yellowish white shining scales. Tasteless. Assumes the appearance of gum in cold water. Becomes opaque and insoluble when boiled with water. The solution of albumen in cold water is coagulated by boiling.

Dilute nitric acid precipitates it.

Perchloride of mercury (corrosive sublimate) causes a white precipitate.

186. The commonest neutral organic substances which are not dissolved by cold water,* but dissolve in boiling water are—

Gelatine	Dextrine or British gum
Soap	Starch
Gum	Oxalate and nitrate of urea.

187. *Starch*, $C_6H_{10}O_5$, identified.

White. Tasteless. Forms a paste when boiled with a

---

* It must be remembered that this is not a very exacting distinction; thus, at the ordinary temperature of the laboratory, dextrine and the oxalate and nitrate of urea dissolve pretty readily.

small quantity of water; even with much water, yields a turbid solution.

Boiled for some time with dilute hydrochloric acid, the solution becomes thinner, and answers to the tests for grape-sugar (180).

Solution of iodine in water, added to the (cold) solution of starch, gives a fine blue color which disappears on boiling, and returns as the liquid cools. Potash destroys the blue color.

188. *Dextrine*, $C_6H_{10}O_5$, identified.

Yellowish. Tasteless. Behaves like gum with cold water.

Boiled for some time with dilute hydrochloric acid, the solution answers to the tests for grape-sugar (180).

Solution of iodine does not give the blue color.

Tribasic acetate of lead (or ammoniacal acetate of lead) does not precipitate the solution of dextrine, which is thereby distinguished from ordinary gum.

Solution of tannic acid does not precipitate dextrine, which thus differs from starch.

189. *Gum*, $C_{12}H_{22}O_{11}$, identified.

Answers to the same tests as dextrine, but its solution is precipitated by tribasic acetate (or ammonia calacetate) of lead. Gum arabic contains lime, which may be detected by oxalate of ammonia.

190. *Gelatine* $C_{41}H_{67}N_{13}O_{16}$, identified.

Tasteless. Its solution in hot water sets to a jelly on cooling. Perchloride of mercury (corrosive sublimate) precipitates it. Tannic acid precipitates it.

Evolves very offensive alkaline vapor when heated in the dry state.

191. *Soap* may be identified by the directions given at pp. 85 and 90.

their aqueous solutions on cooling. The nitrate evolves a remarkably pungent smell when heated in a small tube.

The oxalic and nitric acids may be detected by the tests given in Table H.

The urea may be detected by mercuric nitrate (182), a little potash being first added.

By adding carbonate of baryta to the aqueous solution, evaporating to dryness on a water-bath (231), and treating the residue with alcohol, the urea is dissolved, and may be obtained, by evaporation, in crystals which can be identified as in (182).

If a solution of nitrate or oxalate of urea be mixed with excess of silver nitrate, filtered if necessary, sodium carbonate added till a slight permanent precipitate appears, boiled, filtered, and boiled down to a small bulk in a test-tube, the liquid, on cooling, deposits silver cyanate which may be identified as at (182).

193. The commonest of these organic substances which are insoluble in water, but dissolve in hot alcohol, are—

Stearine	Stearic acid
Palmitic acid	Cholesterine
Rosin	Naphthaline.

194. *Stearine*, $C_{57}H_{110}O_6$, identified.

White. Crystalline. Fuses in water heated to 160° F. (71° C.).* (See 194a.)

Dissolves in boiling alcohol, but not with extreme facility; deposited again on cooling.

Boiled with potash, forms at first a milky *emulsion*, and gradually dissolves, yielding a solution of soap, p. 85. Heated on platinum or porcelain, evolves the characteristic pungent fumes of *acroleine*, and burns with a luminous flame.

---

* One modification of stearine fuses at 131° F. (55° C.)

194a. *To ascertain the exact temperature at which a substance fuses,* draw out a piece of glass tube (Fig. 15, p. 34) to a fine tapering point, and introduce a fragment of the substance to be tested. Warm the tube gently so that the melted substance may run into the point, where it will become opaque on cooling. The tube is now plunged, together with a thermometer, into a beaker (Fig. 55) of water slowly heated over a lamp, and the temperature is noted at which the point becomes transparent, in consequence of the fusion of the substance.

195. *Steric Acid,* $HC_{18}H_{35}O_2$, identified.
White. Crystalline. Fuses in water heated to 160° F. (71° C.). (See 194a.) Dissolves very easily in boiling alcohol; the solution reddens litmus. Potash speedily dissolves it, yielding a solution of soap (191). When heated, it does not evolve the pungent fumes of acroleine. It burns with a luminous flame.

196. *Palmitic Acid,* $HC_{16}H_{31}O_2$, identified.
White. Crystalline. Fuses in water heated to 144° F. (62° C.). (See 194a.) In other respects resembles stearic acid.

197. *Cholesterine,* $C_{26}H_{44}O$, identified.
Transparent tabular crystals. Infusible even in boiling water. Dissolves easily in boiling alcohol, and crystallizes out in rhombic plates. Not changed by boiling with potash.
Fuses easily when heated to 293° F., and passes off in vapor, which burns with a luminous flame.
Moistened with strong nitric acid, and evaporated to dryness, gives a yellow residue changed red by ammonia.
Moistened with strong hydrochloric acid, a little perchloride of iron added, and evaporated to dryness, assumes a violet blue color.

198. *Rosin* identified.
Semi-transparent yellow or brown solid, smelling of tur-

pentine. Dissolves in boiling alcohol; the solution deposits small crystals of (sylvic acid, $HC_{20}H_{29}O_2$,) as it cools.

Dissolves in boiling potash, and is reprecipitated in soft solid flakes by hydrochloric acid. Fuses easily at (260° or 270° F.) and burns with a very smoky flame.

199. *Naphthaline*, $C_{10}H_8$, identified.

Transparent flaky crystals smelling strongly of coal-gas.

Fuses in hot water at 174° F. (79° C.) (See 194a.)

Burns with a very smoky flame. Not changed by boiling with potash. Heated in a dry test-tube, sublimes in crystals.

200. The inflammability of ether and the readiness with which it is converted into vapor, render it necessary to be very careful in using it.

Never bring a bottle of ether within two or three feet of a flame. Replace the stopper immediately.

Fig. 55.

In boiling a substance with ether, do not employ a flame, but place the test-tube in some hot water, and close the orifice lightly with the finger in order to restrain the escape of vapor (fig. 55).

201. The commonest organic substances which are insoluble in water, and sparingly dissolved by alcohol, but soluble in boiling ether, are—

| Palmitine | Wax |
| Spermaceti or cetine | Paraffine. |

202. *Palmitine*, $C_{51}H_{96}O_6$, identified.

White. Crystalline. Fuses in water heated to 145° F. (63° C.)* (See 194a.)

Behaves in other respects like stearine (194), except as to its solubility in alcohol.

---

* One modification of palmatine fuses at 115° F. (46° C.).

203. *Spermaceti*, $C_{32}H_{64}O_2$, identified.

Characteristic pearly crystalline appearance. Fuses in water heated to 120° F. (49° C.). (See 194a.) Boiling with potash does not saponify it. Does not evolve the pungent vapors of acroleine when heated, thus differing from palmitine.

204. *Wax* identified.

White or yellow; not distinctly crystalline. Fuses in water heated to 145° or 150° F. (63° or 66° C.). (See 194a).

Boiled with alcohol, is partly dissolved. When the solution cools, cerotic acid, $C_{27}H_{54}O_2$, crystallizes out, which fuses in water heated to 174° F. (79° C.). The alcoholic solution reddens litmus. When evaporated on a water bath, it leaves ceroleine, a greasy substance of peculiar odor, fusible at 83° F. (28° C.). Wax is little affected by boiling with potash.

205. *Paraffine* identified.

White. Crystalline; resembles spermaceti. Fuses in water heated to 112° F. (44° C.).* (See 194a). Very sparingly soluble in alcohol. Unaffected by boiling with potash. May be distilled with little decomposition, which is not the case with wax.

206. The commonest organic substances which are insoluble in boiling water, alcohol, and ether, and which cannot be distinguished by their *organized* structure,† are—

| Albumen | Caseine. |

Albumen and *Caseine*, in their coagulated or insoluble states, so nearly resemble each other, that no satisfactory test can be given by which they may be distinguished.

They are both white and opaque in the moist state, becoming horny, yellowish, and translucent when dried.

---

* Some specimens of paraffine fuse at 149° F. (65° C.).

† Such substances as fibrin, cellulose, lignine, hair, silk, wool, horn, &c., would always be recognized without having recourse to

Dried albumen and caseine, placed in water, slowly soften, swell, and become white and opaque.

Heated in a tube, they carbonize, swell up, and emit very offensive vapors, which are strongly alkaline to reddened litmus paper.

Strong nitric acid colors them bright orange, and gradually dissolves them when heated.

Solution of nitrate of mercury, prepared by dissolving 2 parts of mercury in 4 parts of nitric acid (sp. gr. 1·40) imparts a bright red color to albumen and caseine (*Millon's test*).

Potash dissolves albumen and caseine, when heated; acetic acid, gradually added to the solution, causes a flocculent precipitate, which is redissolved by an excess of the acid.

If albumen or caseine be boiled with potash, and a few drops of solution of acetate of lead, a dark precipitate of sulphide of lead is produced.

Strong hydrochloric acid slowly dissolves albumen and caseine with the aid of heat, yielding solutions which have a violet color.

LIQUID ORGANIC SUBSTANCES. TABLE Q. 161

2. LIQUID ORGANIC SUBSTANCES.

   A. *The Liquid has a very Distinct Odor.* Examine by Table Q.
   B. *The Liquid has no Powerful or Characteristic Odor.* See (230).

TABLE Q.

207. Examination of a Liquid Organic Substance which has a very Powerful Odor, but which cannot be recognized by the Tests for Organic Acids given at p. 129.

1.	2.	3.
Shake a portion of the liquid with about an equal volume of WATER. If it mixes easily with the water and does not separate again, examine by (208). If it does not mix with the water, or separates again on standing, pass on to column 2.	Add a little POTASH to the water, and shake again. If the liquid mixes with the alkaline solution, and does not separate on standing, examine by (216). If it separate again, pass on to column 3.	Shake a portion of the liquid with DILUTE HYDROCHLORIC ACID. If it dissolves, it is probably *Aniline* (219). If it does not dissolve, examine by (220).

NOTES TO TABLE Q.

208. The commonest liquid organic substances (not distinguishable by the preceding Tables) which have a distinct color, and mix easily with water, are—

Alcohol	Wood-spirit (methylic alcohol)
Aldehyde	Nicotine
Aceton	Butyric acid.

209. *Alcohol*, C_2H_6O, or *Spirit of Wine* (which is a mixture of alcohol and water), may often be recognized at once by its odor.

If not too much diluted, it inflames readily, and burns with a pale flame.

If much water be present, it may be separated either by carbonate of potash (226), or by distillation (227).

When alcohol, even in a diluted state, is mixed with enough chromate or bichromate of potash to color it distinctly, a little hydrochloric acid added, and heat applied, the red color of the solution is changed to green, in consequence of the reduction of the chromic acid to chromic oxide by the deoxidizing effect of the alcohol, a part of which is converted into aldehyde, distinguishable by its peculiar odor.

By heating alcohol with some strong sulphuric acid, and an acetate (either acetate of potash, soda, or lead), the very agreeable odor of acetic ether is developed.

210. *Methylated-spirit* (a mixture of spirit of wine with wood-spirit) may be distinguished from pure spirit of wine by its odor, and by the brown red color which it assumes when mixed with strong sulphuric acid.

211. *Wood Naphtha*, CH_4O (pyroligneous ether, pyroxylic spirit), is not commonly met with in commerce in a pure state, in which form it bears much resemblance to ordinary alcohol. The ordinary wood-naphtha has a yellowish color and a peculiar nauseous odor. When mixed with water, it becomes turbid, from the separation of certain oily impurities.

Wood-naphtha burns with a pale flame, resembling that of alcohol.

Potash immediately imparts a brown color to wood-naphtha, an effect not produced with alcohol until some time has elapsed.

212. *Acetone*, C_3H_6O (wood-spirit), may be recognized by its peculiar odor (which may be ascertained by heating solid acetate of lead in a small tube). It differs also from alcohol and wood-naphtha by burning with a very luminous flame.

213. *Aldehyde*, C_2H_4O, has a very peculiar acrid apple-like smell which affects the eyes. When exposed to the air it passes off in vapor much more readily than alcohol, wood-naphtha, or acetone, first becoming acid from absorption of oxygen.

If aldehyde be added to nitrate of silver mixed with a very little ammonia, the metal is reduced on the application of heat, and forms a mirror-like coating upon the side of the tube. Potash imparts a brown color to aldehyde. Aldehyde is very inflammable, and burns with a pale flame.

214. *Nicotine*, $C_{10}H_{14}N_2$, is an oily liquid, tinged brown by exposure to air, and having a powerful odor of tobacco.

Its aqueous solution is strongly alkaline to test-papers. When the aqueous solution is acidulated with hydrochloric acid, mixed with bichloride of platinum, and allowed to stand, it deposits a precipitate composed of very distinct prismatic crystals.

Nicotine is inflammable, and burns with a smoky flame.

215. *Butyric Acid* $HC_4H_7O_2$, is a colorless liquid, having a most powerful smell of rancid butter. It is somewhat lighter than water, in which it dissolves when shaken. If strong hydrochloric acid be added to the aqueous solution, the butyric acid separates again as an oil upon the surface.

When butyric acid is shaken with alcohol and oil of vitriol, butyric ether is formed, which is recognized by its odor of pine-apple.

216. The commonest liquid organic substances (not distinguishable by the preceding Tables) which have a distinct odor, do not mix easily with water, but are miscible with potash, are—

Carbolic acid (in its liquid form) | Valerianic acid.

217. *Liquid Carbolic Acid* is usually met with as a brownish or brown liquid, having a powerful smell of tar. When poured into water, it sinks to the bottom.

It may be further examined as at (176).

218. *Valerianic Acid*, $HC_5H_9O_2$, is a colorless oily liquid, which floats upon water, and has a powerful odor resembling that of valerian root.

219. *Aniline* C_6H_7N, is usually met with as a yellowish or brown oily liquid, having a strong smell recalling that of ammonia. It sinks in water.

Solution of chloride of lime added in excess to a drop of aniline shaken with water produces an intense purple color. If toluidine be present, as is generally the case with commercial aniline, the purple color passes into brown; but if the mixture be shaken with ether, the latter will rise to the surface, carrying a red-brown coloring matter with it, and leaving the solution of a fine blue color. Deal is stained yellow by aniline.

Oxalic acid combines with aniline to form a sparingly soluble oxalate.

Corrosive sublimate (mercuric chloride) in the solid form, heated with aniline, converts it into a dark purple mass which yields a purple-red solution in alcohol.

220. The commonest liquid organic substances (not distinguishable by the preceding Table) which have a distinct odor and do not mix easily with water, potash, or hydrochloric acid, are—

Ether	Oil of bitter almonds
Chloroform	Nitrobenzole
Benzole	Bisulphide of carbon.

221. *Ether*, $C_4H_{10}O$, may be identified almost with certainty by its odor. It is colorless, very easily inflammable, and burns with a bright flame. Ether very easily passes off in vapor when exposed to the air, so that when the mouth of a test-tube which contains ether is applied to a light the vapor takes fire, and burns at the mouth of the tube if the latter be slightly inclined.

Ether boils at a temperature (94°.8 F., 35° C.) which feels scarcely warm to the hand.

Oil or fat of any kind dissolves very easily in ether.

222. *Chloroform*, $CHCl_3$, is a colorless, very fragrant liquid, which sinks in water (sp. gr. 1.5).

It easily escapes in vapor when exposed to air, and boils at 142° F. (61° C.).

Chloroform dissolves India-rubber with great facility.

When chloroform is gently heated with a solution of hydrate of potash in alcohol it yields chloride of potassium and formiate of potash. The former may be recognized by the white precipitate with nitrate of silver, insoluble in nitric acid, and the latter by neutralizing the alkaline liquid with dilute sulphuric acid, adding an excess of nitrate of silver, decanting the liquid from the precipitate, and gently heating it, when metallic silver will be separated as a dark precipitate, either at once or on adding a drop or two of ammonia.

The *tests for ascertaining the purity of the chloroform* employed in surgical operations, are the following:—

It should be quite free from any odor of chlorine.

When shaken with water, the solution should not redden blue litmus paper, or produce any turbidity with nitrate of silver.

When shaken with oil of vitriol, the mixture should remain colorless.

On evaporating a little chloroform on the hand it should not leave any unpleasant odor.

233. *Oil of Bitter Almonds*, $C_7H_6O_2$, has a very characteristic smell, a yellowish color, and sinks in water.

When heated with solid hydrate of potash it yields benzoate of potash. If the cool mass be dissolved in water, and hydrochloric acid added to the solution, benzoic acid is precipitated (154).

As sold in the shops, the oil of bitter almonds is often dissolved in spirit of wine, from which it is separated on adding water.

224. *Nitrobenzole*, $C_6H_5NO_2$, or *Essence of Mirbane*, much resembles oil of bitter almonds in appearance and odor, but may be easily distinguished from it by converting it into analyne.

For this purpose the nitrobenzole is dissolved in alcohol, some hydrochloric acid added, and a fragment of granulated zinc. When the evolution of hydrogen has nearly ceased, the liquid is mixed with excess of potash and shaken with ether, which dissolves the aniline. When the ether has risen to the surface, it is poured off into a small dish and allowed to evaporate spontaneously, when the aniline will remain, and may be identified as at (219).

225. *Benzole*, C_6H_6, or *Benzine*, or *Benzene*, is a colorless liquid which smells strongly of coal-gas.

It floats on water, is very inflammable, and burns with a luminous smoky flame.

When added, drop by drop, to the strongest nitric acid, benzole is dissolved, with evolution of much heat and red fume, to a red liquid, and if this be poured into a large volume of water, a heavy oil is separated, which is nitrobenzole, and may be identified as described above (224).

Bisulphide of Carbon, CS_2, is a colorless or yellowish liquid, which sinks in water and has a most offensive smell.* It is extremely inflammable, and burns with a blue flame, emitting a powerful odor of sulphurous acid. If a few drops be placed in a watch-glass, and blown upon, it will evaporate very rapidly, condensing the moisture upon the glass into white hoar-frost, and freezing a part of the bisulphide to a white crystalline mass.

226. *Removal of water from alcohol by carbonate of potash.*—Pour the liquid into a large test-tube, or a draught-bottle furnished with a good cork, so that the tube or bottle may be about half-filled. Introduce dried powdered carbonate of potash, in small portions, shaking well after each addition, as long as it dissolves in the liquid. If alcohol be present, it will form a separate layer upon the surface of the solution of carbonate of potash in water. Pour off this layer carefully into another tube, and dip a glass-rod in it to test its inflammability. It may then be examined by other tests for alcohol.

227. *Separation of alcohol and water by distillation.*— To separate alcohol from water by distillation, the mixture must be maintained for some time at a temperature below 212° F. (100° C.), when the alcohol will rise in vapor much more readily than the water, and if the first portions of vapor be condensed and collected in another vessel, they will be found to contain the chief part of the alcohol.

---

* Purified bisulphide of carbon has not an offensive odor.

DISTILLATION.—The best form of apparatus for distillation is that represented in fig. 57, where *a* represents a *Retort*, through the *tubulus* (*b*) of which a thermometer* (*c*) is fixed by means of a perforated cork (228) so that the bulb of the thermometer nearly touches the bottom of the retort. The neck of the retort is thrust into the inner tube (*d*) of a *Liebig's condenser*, through the outer tube (*e*) of which a constant flow of water is maintained by means of the pipes (*f*) (which comes from the tap) and (*g*) (which runs into the sink). These pipes are vulcanized India-rubber, and (*f*) is slipped on to a piece of gas-pipe or glass tube (*h*) bent into a hook, so as to hang upon the funnel (*i*). The joint (*k*), where the retort neck is fitted into the condensing tube, is secured by a

FIG. 56. FIG. 57.

Distillation.

tight bandage made by warming a piece of sheet india-rubber about four inches long and one broad, securing one end of it with the thumb over the joint, and stretching it very considerably whilst binding it round the tubes (fig. 58). The condensed liquid drops into the bottle (*m*), which may be changed when necessary without disturbing the apparatus.

Thermo-
meter.

DISTILLATION.

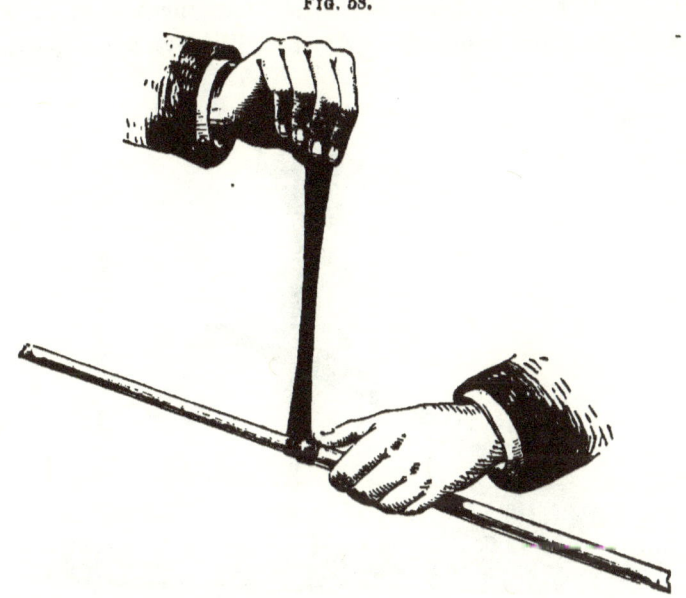

Fig. 58.

Heat is gradually applied to the retort, either by a rose gas-burner (fig. 59), or a plain ring burner (fig. 60), or an Argand burner with a chimney (fig. 32, p. 94).

Since alcohol boils at 173° F. (78°·3 C.), a rough estimate of the proportion of alcohol present may be formed from the quantity of liquid which distils over at a few degrees above that temperature, and the distillation may be stopped when the thermometer approaches 212° F. (100° C.) and the taste and smell of the liquid distilling over (*distillate*) indicate the presence of very little alcohol.

Fig. 59.

Rose Burner.

Where such an apparatus as that just described is not to be obtained, some simpler contrivance must be substituted for it.

A plain retort (*a*, fig. 60) with a long neck may be employed, and any common bottle (*b*) will serve for a receiver.

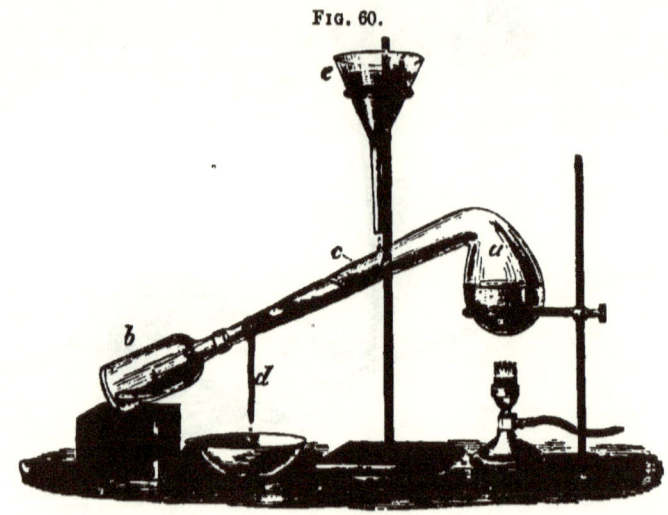

Fig. 60.

Distillation.

To promote condensation, a long strip of filter-paper (*c*) may be wetted and wrapped smoothly round the neck, a string of wet tow (*d*) being passed twice round the neck at the lower edge of the paper, and twisted tightly into a tail to carry off the water, which may either be gently poured from time to time upon the upper part of the paper, or allowed to trickle slowly from a funnel (*e*), the neck of which is partly stopped with tow. A tube funnel (fig. 61) is employed for introducing the liquid into the retort without soiling the neck.

Fig. 61.

Tube funnel.

A flask with a bent tube (229), tightly fitted into it with a perforated cork (228), may be employed instead of a retort. One limb of this tube may be 20 or 30 inches long, to insure condensation, or it may be adapted, either by a perforated cork or a caoutchouc bandage, to a wider tube of considerable length (fig. 62). A convenient support for this tube is made by fixing a perforated bung into the ring of a retort-stand turned round into the required position.

FIG. 62.

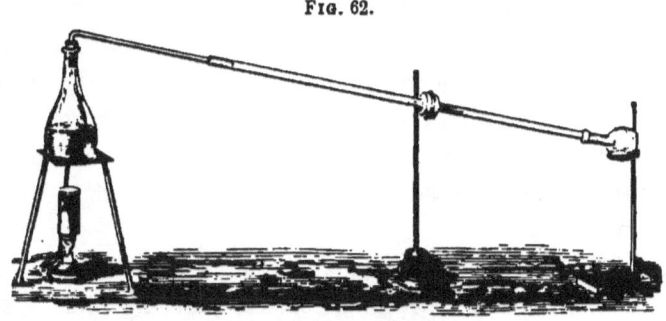

Distillation in a flask.

228. *To perforate corks.*—Smooth cylindrical holes are made in corks with *rat's-tail files* (fig. 63), beginning with a

FIG. 63.

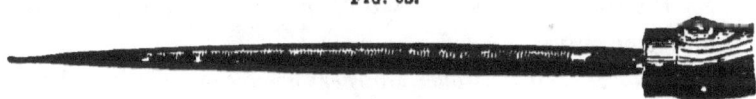

Rat's-tail file.

small size, and employing the larger files as may be necessary. Corks should always be kept on the points of the files when not in use, as the steel is very brittle.

A set of brass cork borers of various sizes (fig. 65) will save much time. They are made to slip into each other, and are provided with a steel rod (fig. 64) which serves as a handle and for thrusting out the cylinders of cork punched by the borers. A cork-borer is selected of somewhat less diameter than the tube for which the hole is to be bored, and the rod is thrust through the holes in the head of the borer. The cork is held firmly against the wall or the

FIG. 64.

FIG. 65.

Set of cork-borers.

FIG. 66.

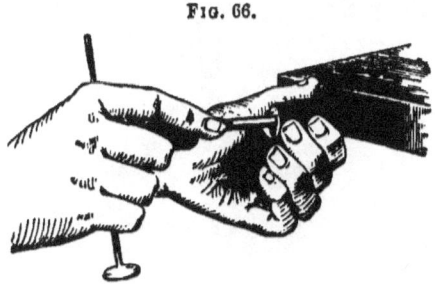

and the borer worked straight into it, like a gimlet, until it is about half way through the cork. The borer being withdrawn, and cleared, if necessary, with the rod, the cork is reversed, and bored in the opposite direction, so that the two holes may meet in the centre, and form a perfectly smooth cylindrical passage, which is very carefully enlarged with a rat's-tail file until it is just large enough to receive the tube, which should pass through it with considerable friction.

In *fitting corks air-tight* they should be carefully selected as free from flaws as possible, especially at the ends. The cork should be somewhat too large to enter the mouth of the vessel until it has been softened by rolling it heavily on the table with the palm of the hand, or, in the case of large corks, under the sole of the boot. Corks are always to be preferred to bungs or shives.

Vulcanized India-rubber stoppers are often substituted for corks, and are decidedly preferable in a great many cases. They may be perforated with the cork-borers described above, which should be dipped in spirit of wine.

229. *To bend glass tubes.*—Small tubing may be bent either in the flame of a spirit-lamp, or in the upper part of a somewhat flaring gas-flame (fig. 67). The tube should be slowly rotated, and moved to and fro in the flame until soft enough to be bent, which should be effected by a gentle equal pressure with both hands, care being taken so to regulate the soft-

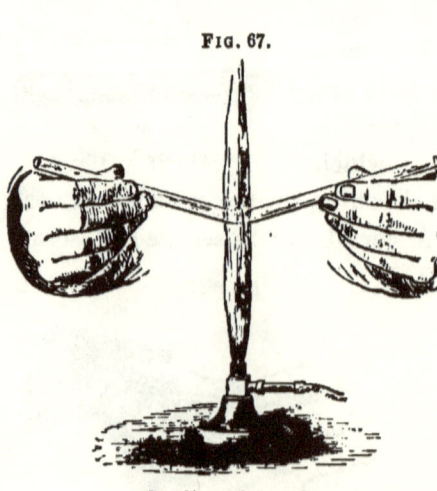

FIG. 67.

Bending glass tube.

ening of the glass as to obtain a nice curve (fig. 68) instead of a sharp angle (fig. 69). Any soot which has been deposited from the flame may be wiped off with paper when the tube is cool.

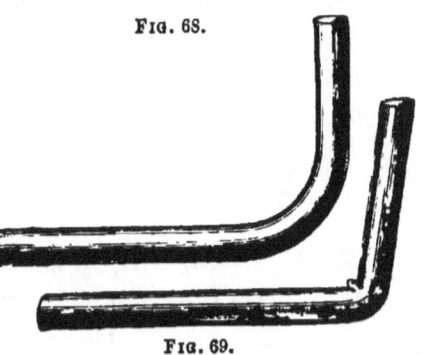

Fig. 68.

Fig. 69.

Large tubing is more difficult to bend, and it is often necessary to employ a blowpipe flame. The bend must be annealed by withdrawing it very gradually from the heat.

The gas blowpipe represented in fig. 70 is very convenient for such purposes, especially if connected with a double-action bellows worked by the foot.

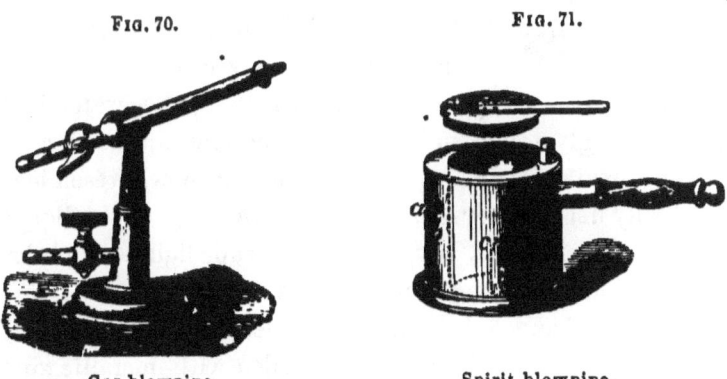

Fig. 70.   Fig. 71.

Gas blowpipe.   Spirit blowpipe.

Where gas is not to be had, a spirit blowpipe-lamp is sometimes used. That represented in fig. 71 answers the purpose very well. A small quantity of spirit (either methylated spirit of wine or wood-naphtha) burnt inside the vessel a vaporizes the spirit in the space b between the walls; the vapor issuing from the jet c, burns with a powerful flame. These lamps are not free from danger in consequence of a

jet, when the operator should at once place the cover on the lamp, and thus extinguish the flame.

230. *If the liquid has no powerful or characteristic odor it may be—*

        Glycerine | Lactic acid
        Oleine | Oleic acid.

231. *Glycerine*, $C_3H_8O_3$, is a syrupy liquid which has an intensely sweet taste and mixes readily with water.

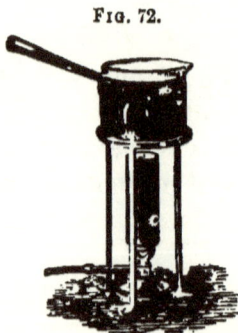

Fig. 72.

Heated sharply on a knife-blade or a piece of platinum foil, it burns with a luminous flame. No residue is left.

When heated with strong sulphuric acid, it blackens and evolves very pungent vapors of acroleine, which strongly affect the eyes.

The same substance is produced by heating a little bisulphate of potash moistened with glycerine.

Placed in an evaporating dish heated upon a water-bath (fig. 72), glycerine suffers no perceptible diminution or change, whilst ordinary syrup, which it much resembles, gradually deposits crystals of sugar at the edge of the liquid.

232. *Lactic acid*, $HC_3H_5O_3$, is a syrupy liquid which has a strong acid taste and readily mixes with water.

It is not changed by heating on the water-bath (fig. 72). If it be diluted with water, and boiled with metallic zinc, the solution, on cooling, deposits crystalline crusts of lactate of zinc.

When lactic acid is heated in a retort, several products are distilled over, and among them a crystalline solid known as lactide, $C_3H_4O_2$, which is soluble in hot strong alcohol, and is deposited in transparent flat prismatic crystals on cooling.

233. *Oleine*, $C_{57}H_{104}O_6$, and *oleic acid*, $HC_{18}H_{33}O_2$, are colorless or yellow oils which do not mix with water, but float upon its surface.

Alcohol dissolves oleic acid more readily than oleine.
Ether dissolves them both very readily.
Solution of potash dissolves oleic acid more easily than it dissolves oleine.
Oleine when strongly heated evolves the odor of acroleine, which is not produced when oleic acid is heated.
If a test-tube containing oleic acid is placed in melting ice the oleic acid solidifies to a mass of needle-like crystals, but oleine remains liquid. When oleic acid has been kept for some time in contact with air it acquires a brown color and an acid reaction. It does not then solidify at the melting-point of ice.

234. *Examples for Practice in Exercise* **X.**—Since it is only by a careful study of individual organic substances that the analyst can learn to identify them with certainty, the student is recommended to examine as many of the substances mentioned in this Exercise as he is able to procure. A list of them is subjoined.

Acetone	Ether	Palmitic acid
Albumen (white of egg)	Gelatine	Palmitine
	Glycerine	Paraffine
Alcohol	Grape-sugar	Picric or carbazotic acid
Aldehyde	Gum arabic	
Aniline	Indigo	Pyrogallic acid
Benzole	Lactic acid	Rosin
Butyric acid	Methylated alcohol	Salicine
Cane-sugar	Milk-sugar	Soap
Caramel	Naphthaline	Spermaceti
Carbolic acid	Nicotine	Starch
Caseine (curd of milk)	Nitrate of urea	Stearic acid
	Nitrobenzole	Stearine
Chloral hydrate	Oil of bitter almonds	Urea
Chloroform	Oleic acid	Valerianic acid
Cholesterine	Oleine	Wax
Dextrine	Oxalate of urea	Wood-naphtha.

EXERCISE XI.

235. EXAMINATION OF A SOLID ORGANIC SUBSTANCE ABOUT WHICH NOTHING IS KNOWN BUT THAT IT IS A SINGLE SUBSTANCE AND NOT A MIXTURE.

The analyst is recommended to mark off each substance from the subjoined list as it is excluded by his experiments, and in this way to reduce the number of possible substances within very narrow limits :—

Acetates
Albumen
Aniline salts
Benzoates
Benzoic acid
Brucine
Caffeine
Cane-sugar
Caramel
Carbolic acid
Caseine
Cholesterine
Cinchonine
" sulphate
Citrates
Citric acid
Cyanides
Dextrine
Ferridcyanides
Ferrocyanides
Gallic acid
Gelatine
Grape-sugar
Gum
Hippuric acid
Indigo
Malic acid
Meconic acid
Milk-sugar
Morphine
" acetate
" hydrochlorate
" meconate
Naphthaline

Narcotine
Nitroprussides
Oxalates
Oxalic acid
Palmitic acid
Palmitine
Paraffine
Picric acid
Prussian blue
Pyrogallic acid
Quinine
" and iron, citrate
" sulphate
Rosin
Salicine
Soap
Spermaceti
Starch
Stearic acid
Stearine
Strychnine
Succinic acid
Sugar
Sulphocyanides
Tannic acid
Tartaric acid
Tartrates
Urates
Urea
" nitrate
" oxalate
Uric acid
Wax.

A. Heat the substance gradually in a small glass tube (17).

(a) *It fuses easily, becoming perfectly liquid.*

EXCLUDES.

| Albumen | Dextrine | Urates |
| Caseine | Starch | Uric acid. |

(b) *It passes off in vapor*
   (with or without previous fusion)
   and leaves no black (carbonaceous) residue.

EXCLUDES

Albumen	Morphine and its salts
Caramel	Quinine and its salts
Caseine	Salicine
Cinchonine and its salts	Starch
Citric acid and citrates	Strychnine and its salts
Dextrine	Sugar
Gallic acid	Tannic acid
Gelatine	Tartaric acid and tartrates
Gum	Uric acid and urates.

(c) *It passes off in vapor, leaving no residue whatever.*

EXCLUDES

all the substances in the above list, and, in addition, all metals except mercury, arsenic, and ammonium.

(d) *It emits vapors which have the odor of Ammonia, and change red litmus paper to blue.*

PROBABLE PRESENCE OF

Ammonia, combined with an organic acid	Gelatine
Urea	Albumen
Uric acid	Caseine
A ferrocyanide	Morphine
A cyanide	Quinine
A sulphocyanide	Cinchonine
A ferridcyanide	Strychnine
	Aniline.

The analyst should mark off such of these as have been excluded by the previous experiments.

178   ORGANIC SUBSTANCES EXAMINED.

If the evolution of ammonia is abundant and unattended by any carbonization, either urea or one of its salts is probably present (182, 192).

B. Heat the substance on a piece of porcelain, and continue the heat until no further change is perceptible, directing the outer blowpipe-flame upon it if necessary, to burn off the carbon.

(*a*) *No residue is left.*

Excludes

all metals except mercury, arsenic, and ammonium.

(*b*) *A residue is left, which is strongly alkaline to moistened red litmus paper.*

Probable Presence of

an Organic Acid, in combination with Potash, Soda, Baryta, Strontia, or Lime.

(See Table K.)

C. Shake a little of the substance with cold water,* in a test-tube.

(*a*) *It dissolves easily.*

Excludes

Albumen	Palmitine
Benzoic acid (free)	Paraffine
Brucine	Quinine (free)
Caffeine	Rosin
Caseine	Salicine
Cholesterine	Spermaceti
Cinchonine (free)	Starch
Dextrine (?)	Stearic acid (free)
Gallic acid (free)	Stearine
Gelatine	Strychnine (free)
Gum	Urates
Hippuric acid (free)	Urea nitrate (?)
Morphine (free)	" oxalate (?)
Narcotine	Uric acid.
Palmitic acid (free)	

* See foot-note on page 154.

(b) *It does not dissolve.*
Boil it with the water; should it dissolve, this

Excludes

Albumen	Palmitine	Stearic acid (free)
Caseine	Paraffine	Stearine
Palmitic acid (free)	Spermaceti	Wax.

D. Very cautiously taste a particle of the substance.
    (a) *Its taste is acid.*
        Examine for an organic acid by Table K, and for an inorganic acid by Table G.
    (b) *Its taste is bitter;* pass on to F.

E. *If the substance does not dissolve readily in water, but dissolves on adding a little potash, and is precipitated by the addition of hydrochloric acid—*

Examine especially for

Benzoic acid (154) | Hippuric acid (156)
              Uric acid (161)

F. Dissolve a little of the substance in water, or in a little dilute hydrochloric acid, and test with a solution of iodine in iodide of potassium.

If a *brown precipitate* is obtained, examine for an alkaloid by Table O.

If a *yellow precipitate* is produced, examine for acetate of lead (p. 29).

If a *blue precepitate* is produced, examine for starch (187).

If the iodine-solution is *bleached*, examine for a cyanide (99).

EXERCISE XII.

236. EXAMINATION OF A LIQUID ORGANIC SUBSTANCE ABOUT WHICH NOTHING IS KNOWN, BUT THAT IT IS A SINGLE SUBSTANCE AND NOT A MIXTURE.

A. Ascertain whether it has any odor or taste characteristic of

Acetic acid (149)	Bitter almond oil (223)	Glycerine (231)
Acetone (212)	Butyric acid (215)	Nicotine (214)
Alcohol (209)	Carbolic acid (217)	Nitrobenzole (224)
Aldehyde (213)	Chloroform (222)	Sugar (179–181)
Aniline (219)	Ether (221)	Valerianic acid (218)
Benzole (225)	Fomic acid (150)	Wood-naphtha (211).

B. Evaporate a little of the liquid in a porcelain dish, carefully observing any odor which may be developed, and stopping the evaporation as soon as the liquid has disappeared.*

    (*a*) *If no residue is left—*

        The liquid probably contains one of the above-mentioned substances recognizable by their odor, or possibly glycerine (231), or lactic acid (232).

    (*b*) *If an oily inflammable residue is left—*

        Examine especially for—

        Oleine (233) | Oleic acid (233).

    (*c*) *If a solid residue is left—*

        Evaporate a large quantity of the liquid and examine the residue according to (235).

C. Whilst the evaporation is proceeding, the analyst should examine another portion of the liquid by Tables L, O, Q.

---

* Should time permit it, it is well to evaporate over a steam-bath.

EXERCISE XIII.

237. EXAMINATION OF A SOLID SUBSTANCE OF WHICH NOTHING IS KNOWN, BUT THAT IT IS A SINGLE SUBSTANCE, AND NOT A MIXTURE.

A. Heat a little of the substance on a piece of porcelain, and observe whether there is any carbonization or peculiar odor to indicate the presence of organic matter.*

B. Heat another portion of the substance with strong sulphuric acid, and observe whether any carbonization indicative of organic matter takes place.

   (a) *If organic matter is detected*, the substance must be examined, according to (235).

   (b) *If no organic matter is detected*, the substance may be examined, according to Tables A to I, or by the blowpipe, according to Tables R to Z.

EXERCISE XIV.

238. EXAMINATION OF A LIQUID OF WHICH NOTHING IS KNOWN BUT THAT IT IS A SOLUTION OF A SINGLE SUBSTANCE, AND NOT OF A MIXTURE.

A. Observe its smell, taste, and action upon test-papers (18).

B. Evaporate a little on a slip of glass (p. 20).

   (a) *If no residue is left*, and the liquid is destitute of color, smell, taste, and action on test-papers, it is water only.

   (b) *If a residue is left*, or if the conclusion is doubtful, evaporate a larger quantity of the liquid in a porcelain dish (84), and if there is any residue,

---

* Sulphur and phosphorus would of course be recognized in this experiment.

examine it as directed for an unknown solid substance (237). Carefully observe whether any odor is evolved during the evaporation.

Whilst the evaporation is proceeding, the analyst may examine another portion of the solution by Tables A and H.

(c) *If no residue is left on evaporation, and the liquid is acid to test-papers,* examine for

Sulphuric acid (102)	Butyric acid (215)
Hydrochloric acid (105)	Hydriodic acid (93)
Nitric acid (109)	Hydrofluoric acid (89)
Acetic acid (149)	Hydrofluosilicic acid (64)
Sulphurous acid (100)	Carbonic acid (95)
Chloric acid (90)	Hydrocyanic acid (99)
Formic acid (150)	Hydrosulphuric acid (97)
Lactic acid (232)	Valerianic acid (218)

Oleic acid (233).

(d) *If no residue is left on evaporation, and the liquid is alkaline to test-papers,* examine for

Ammonia (75) | Aniline (219) | Nicotine (214).

(e) *If no residue is left on evaporation, and the liquid is neither acid nor alkaline,* examine especially for

Alcohol (209)	Ether (221)
Methylic alcohol (211)	Chloroform (222)
Acetone (212)	Bitter almond oil (223)
Aldehyde (213)	Nitrobenzole (224)
Phenole (carbolic acid) (217)	Benzole (225)

Glycerine (231).

When no clue can be obtained, the analyst must *carefully go through every step of the analytical processes, commencing at Table A.*

EXERCISE XV.

239. Examination of an Organic Substance which is known to be included in the following List:—*

Organic Acids.	Organic Alkaloids.	Salts of Organic Acids with Inorganic Bases.	Salts of Organic Alkaloids with Inorganic Acids.
Acetic	Cinchonine	Acetate of potash	Hydrochlorate of morphine
Benzoic	Morphine	" " lead	Nitrate of urea
Citric	Quinine	" " soda	Sulphate of cinchonine
Gallic	Strychnine	Bitartrate of potash	" " quinine, etc.
Hippuric	Urea.	Cyanide of mercury	
Hydrocyanic		" " potassium	**Salts of Organic Acids with Organic Alkaloids.**
Hydrosulphocyanic	*Neutral Organic Substances.*	Ferridcyanide of potassium	
Malic		Ferrocyanide of potassium	Acetate of morphine
Meconic	Alcohol	Oxalate of ammonia	Meconate of morphine
Oxalic	Cane-sugar	" " lime	Oxalate of urea, etc.
Succinic	Glycerine	" " potash	
Tannic	Grape-sugar	Prussian blue	
Tartaric	Starch.	Sulphocyanide of potassium	
Uric.		Urate of ammonia	
		" " soda, etc.	

* Given for the 1st M.B. Examination of the University of London (Calendar, 1871).

I. The Substance is Solid.

A. *Heat the substance in a small tube.*

(a) *It evolves ammonia freely, without blackening.*

Examine especially for—

Urea (182) | Oxalate of urea (192) | Oxalate of ammonia.

(b) *It evolves a pungent odor resembling sulphurous acid.*

Examine for Nitrate of Urea (192).

(c) *It evolves an agreeable odor.*

Examine especially for—

Benzoic acid (154) | Hippuric acid (156) | Acetate of lead (149).

(d) *It evolves an odor resembling that of coal tar.*

Examine especially for—

Cinchonine (170)	Sulphate of cinchonine (170)
Quinine (168)	Sulphate of quinine (168)

(e) *It blackens and evolves an odor of burnt wood or sugar.*

Examine especially for—

Tartaric acid (158)	Grape-sugar (180)
Bitartrate of potash (158)	Malic acid (160)
Citric acid (159)	Meconic acid (151)
Starch (187)	Gallic acid (153)
Cane-sugar (179)	Tannic acid (152).

(f) *It blackens and evolves a disagreeable odor of singed animal matter.*

Examine especially for—

Uric acid (161)	Acetate of morphine (165)
Urate of soda (161)	Meconate of morphine (165)
Urate of ammonia (161)	Strychnine (167)
Morphine (165)	Ferrocyanide of potassium (113)
Hydrochlorate of morphine (165)	Ferridcyanide of potassium (p. 84).

(g) *It evolves vapors which provoke violent coughing.*

Examine especially for Succinic Acid (155).

(*h*) *It evolves cyanogen* (known by its odor, and pink flame).

Examine especially for Cyanide of Mercury (99).

(*i*) *It passes off in vapor without any of the above indications.*

Examine especially for—

| Oxalate acid (115). | Oxalate of ammonia (115). |

B. *Heat the substance on a piece of broken porcelain, at first with the ordinary flame, and afterwards with the outer blowpipe-flame.*

(*a*) *A residue is left.*

Moisten it with water, and test with red litmus paper.

If it be *decidedly alkaline,*

Examine for—

Cyanide of potassium (p. 84)	Oxalate of potash (115)
Ferrocyanide of potassium (113)	" lime (115)
Ferridcyanide of potassium (p.84)	Acetate of potash (149)
Sulphocyanide of potassium (p. 101)	" soda (149)
	Urate of soda (161)
Bitartrate of potash (158)	Tartar emetic (p. 49).

If the residue is yellow or red,

Examine especially for Acetate of Lead (149).

(*b*) *No residue is left.* Pass on to C.

C. *Heat the substance with water.*

(*a*) *It does not dissolve in water.*

Add Potash.

If it dissolves in Potash,

Examine especially for*—

Benzoic acid (154)	Bitartrate of potash (158)
Hippuric acid (156)	Urate of soda (161)
Uric acid (161)	Urate of ammonia (161).
Gallic acid (153)	

---

* It is not safe to infer the absence of these when the substance dissolves in water.

If it does not dissolve in potash,
Add Hydrochloric Acid.*
It dissolves.

Examine especially for—

Morphine (165)	Sulphate of quinine (168)
Quinine (168)	" cinchonine (170)
Cinchonine (170)	Oxalate of lime (58).
Strychnine (167)	

(*b*) *It dissolves in water.*

Test the solution with blue and red litmus paper, and, cautiously, by tasting.

I. *The solution is decidedly acid, but not astringent.*

Examine especially for—

Tartaric acid (158)	Malic acid (160)
Oxalic acid (115)	Meconic acid (151)
Citric acid (159)	Hippuric acid (156)
Succinic acid (155)	Bitartrate of potash (158).

II. *The solution is acid and astringent.*

Examine especially for Gallic (153) and Tannic Acids (152).

III. *The solution is alkaline.*

Examine especially for Cyanide of Potassium (99).

IV. *The solution has a sweet taste.*

Examine especially for—

Cane-sugar (179) | Grape-sugar (180) | Acetate of lead (149).

V. *The solution is bitter; pass on to D.*

VI. *The solution has a yellow or green color.*

Examine especially for—

Ferrocyanide and Ferridcyanide of potassium (p. 80).

---

* A blue substance, becoming brown with potash, and again blue with hydrochloric acid, is probably Prussian blue. To confirm, boil with potash, filter, and test the solution with excess of acetic acid, and perchloride of iron. Abundant blue precipitate indicates Prussian blue.

VII. *The solution is not decidedly acid or alkaline, and has no characteristic taste or color.*

Examine especially for—

Oxalate of ammonia (115)	Cyanide of mercury (31)
" potash (115)	Sulphocyanide of potassium
Acetate of potash (149)	(p. 101)
" soda (149)	Urea (182).

D. *Dissolve a little of the substance in water, or in a little dilute hydrochloric acid, and test with a solution of iodine in iodide of potassium.*

If a *brown precipitate* is obtained,

Examine for an alkaloid by Table O.

If a *yellow precipitate* is produced, examine for acetate of lead (p. 29).

If a *blue precipitate* is produced, examine for starch (187).

If the iodine solution is *bleached*, examine for a cyanide (99).

Should no clue have been hitherto obtained, examine the substance for an organic acid by Table K, and for an organic alkaloid by Table O; and, as a last resource, for a metal by Table A, and for an inorganic acid by Table G, in the hope that some light may thus be thrown upon the nature of the substance.

II. THE SUBSTANCE IS A LIQUID.

A. *Ascertain whether it has the odor of*

Acetic acid (149) | Hydrocyanic acid (99) | Alcohol (209).

B. *Evaporate a little in a porcelain dish*, taking care not to continue the heat after the dish is dry, and observing any odor during evaporation.

(*a*) *If no residue is left*, examine for

Acetic acid (149)	Alcohol (209)
Hydrocyanic acid (99)	Glycerine (231).

(*b*) *If a residue is left*, examine the action of heat upon it, drawing inferences according to pp. 184, 185.

C. *Test a small portion of the solution with a solution of iodine in iodide of potassium.*

If a *brown precipitate* is obtained, examine another part of the solution for—

Morphine (165)	Meconate of morphine (165)
Hydrochlorate of morphine (165)	Quinine (168)
	Cinchonine (170)
Acetate of morphine (165)	Strychnine (167).

Yellow precipitate, examine for acetate of lead (p. 29).
Blue precipitate, examine for starch (187).

If the iodine solution is bleached, examine for a cyanide (99).

D. *Test the liquid with blue and red litmus papers.*
    (a) *The liquid is acid.*

        Add, to a small portion, potash, till it is very slightly alkaline, and stir with a glass rod.

        If a precipitate is produced, add acetic acid in excess; should this fail to redissolve it, examine for Oxalate of Lime (58).

        If acetic acid dissolves the precipitate, examine for—

Acetate of lead (149)	Quinine (168)
Morphine (165)	Cinchonine (170)
Strychnine (167).	

    (b) *The liquid is alkaline.*

        Add, to a small portion, acetic acid, till it is slightly acid, and stir with a glass rod.

        If a precipitate is obtained, examine for—

Benzoic acid (154)	Uric acid (161)
Hippuric acid (156)	Tartaric acid (158).

And for the potash, soda, or ammonia holding them in solution (72).

        If no precipitate is obtained, add to another portion hydrochloric acid (conc.) in slight excess, and stir with a glass rod.

Benzoic and hippuric acids would now be precipitated, though they might have escaped precipitation by acetic acid.

If no clue has yet been obtained, examine the substance for an organic acid by Table K, and for an organic alkaloid by Table O; and, as a last resource, for a metal by Table A, and for an inorganic acid by Table G, in the hope that some light may thus be thrown upon the nature of the substance.

EXERCISE XVI. (Seee (282) for Examples for Practice.)
ANALYSIS BY THE BLOWPIPE.
TABLE R.

240. Detection of Metals by the Blowpipe.

1.	2.	3.	4.	5.
Heat a little of the powdered substance on CHARCOAL,* with CARBONATE OF SODA, in the inner flame of the blowpipe (241). Observe whether there is any decided odor, any metallic globule, or any marked deposit on the surface of the charcoal. Compare the result with Table S. If there be no decided result, pass on to column 2.	Take a very little of the powdered substance upon a hot bead of fused BORAX in a loop of platinum wire (243). Expose the bead to the outer blowpipe-flame for a minute or two. Carefully note the color of the hot and cold glass, and expose to the inner flame. Compare the color of the glass with Table T. If no distinct color is imparted to the glass, pass on to column 3.	Moisten a clean platinum wire with HYDROCHLORIC ACID, take upon it a very little of the powdered substance, and expose it to the point of the inner blowpipe-flame, observing whether any distinct color is imparted to the outer flame (244). Compare the result with Table U. If no distinct color (except yellow) is imparted to the flame, pass on to column 4.	Heat a little of the powdered substance on charcoal at the point of the inner blowpipe-flame, until it leaves an infusible residue. Moisten this with a drop or two of NITRATE OF COBALT (245), and again heat very intensely in the point of the inner flame. Blue mass indicates *Aluminium* (247). Green mass indicates *Zinc* (248). Pink mass indicates *Magnesium* (249). If neither color is obtained, pass on to column 5.	Mix a little of the powdered substance with dried CARBONATE OF SODA and a little CHARCOAL, and heat in a small tube closed at one end (246). *Mercury* condenses in minute gray globules on the cooler part of the tube (250). *Arsenic* yields a shining black sublimate (251). *Ammonium* compounds evolve an odor of ammonia (252).

* In examining substances which are entirely unknown, it will be found advantageous to make a preliminary experiment without adding the carbonate of soda (242a).

EXPLANATIONS AND INSTRUCTIONS ON TABLE **R**.

241. *To reduce metals on charcoal before the blowpipe.*—Select a piece of hard thoroughly carbonized charcoal, free from crevices, not less than four inches long and one or two inches in diameter; grind down one of its sides to a flat surface (fig. 73) on the hearthstone. Scoop a very shallow

FIG. 73.

cavity at *a* with the blade of a knife, making it smooth and round. Place in this a grain or two of the substance to be examined, previously reduced to powder (124), and cover it with dry powdered carbonate of soda. Hold the charcoal and the blowpipe in the positions represented in fig. 74. Direct the point of the inner (reducing) flame upon the specimen in the cavity, blowing gently at first lest the powder

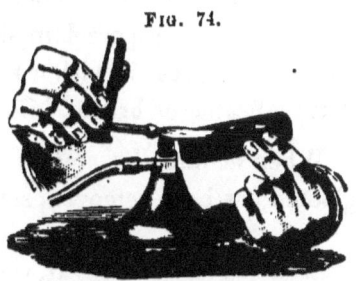

FIG. 74.

Reduction on charcoal.

should be scattered, and allow the outer (oxidizing flame) to play over the flat surface of the charcoal.

Observe very closely the appearances presented by the mass under the influence of heat, especially noticing whether any minute metallic globules are to be seen in the fused substance. If this be the case, try to fuse them together into larger globules.

Should an *infusible mass* be left after the first application of the blowpipe-flame, add more carbonate of soda, and again heat intensely, since the binoxide of tin often requires repeated additions of carbonate of soda to bring it into fusion and reduce it to the metallic state.

Watch the appearance of the mass after withdrawing it from the flame, noting any changes of color which may occur in cooling.

The surface of the charcoal is generally covered, for some distance beyond the cavity, with a deposit or *incrustation* which sometimes consists of a thin white film of ash left after the charcoal has burnt away, and sometimes of a more opaque coating of some metallic oxide formed by the combustion of metallic vapor in passing through the outer flame.

Observe very carefully the color and general appearance of this *incrustation*, comparing the results with Table S.

If any globule of metal is visible, detach it carefully from the fused mass with the point of a knife, place it upon a hard surface, such as a porcelain slab or the bottom of an inverted mortar, and press it with a knife-blade, to ascertain whether it is malleable or brittle. Compare the results with Table S.

When *no metallic globule is visible*, or when the metallic globule has been removed, scrape the mass, together with the particles of charcoal in contact with it, into a small agate mortar (fig. 75), moisten it with one or two drops of water, and grind it into a paste. Stir this paste up with more water, then fill the mortar with water, allow it to rest for a few seconds, in order that any metallic particles may subside, and carefully pour off the water, carrying with it the lighter particles of charcoal and

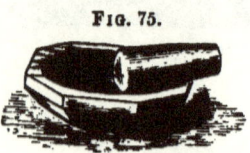

Fig. 75.

Agate mortar.

slag (fig. 76). Repeat this grinding and *levigation* until metallic particles are distinctly visible at the bottom of the mortar, or until the whole has been washed away without showing any metal.

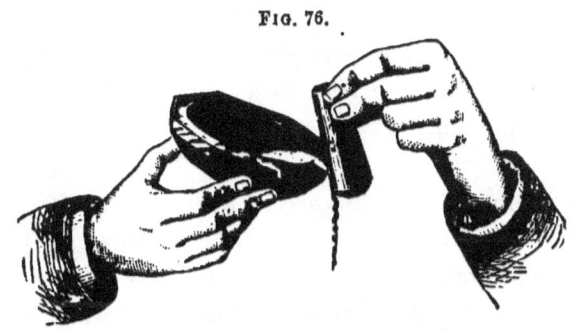

FIG. 76.

Levigation in blowpipe analysis.

The metals which are generally detected in this way, are—

Copper, which gives characteristic red spangles.

Tin, in white silvery spangles of considerable size.

Iron, in gray metallic powder, attracted by the magnet.

242. Fused carbonate of soda is absorbed into the pores of the charcoal, but very frequently a slag is formed which refuses to sink into the charcoal, and remains on the surface.* This is the case with silicate and borate of soda, formed when silicic and boracic acids are present. Sulphide of sodium also generally remains on the surface of the charcoal as a *brown mass*, the formation of which renders it highly probable that the substance under examination is a *sulphide*.

Although most metallic oxides would be reduced to the metallic state by the combined action of the blowpipe-flame and the charcoal support, it is necessary to add carbonate of soda for the following reasons:—

(1) The carbonate of soda removes any acid (silicic acid,

---

* Cyanide of potassium will occasionally assist in getting rid of such slags.

for example) or non-metallic element (such as sulphur) which would hinder the separation of the metal.

(2) By thus forming a slag through which the metal may sink, the re-oxidation, and in some cases the volatilization, of the metal, are in great measure prevented.

(3) Carbonate of soda, strongly heated with charcoal, yields vapor of sodium, which acts as a powerful reducing agent upon metallic compounds.

242a. When the substance is heated by itself on charcoal it may furnish any of the results indicated in Table S, and, in addition, other information may often be obtained more easily than when carbonate of soda has been added.

(1) The substance *passes away entirely in the form of vapor.* Probably mercury, arsenic, or ammonium is present. See Table R, col. 5.

(2) An *infusible mass remains* upon the charcoal. Probably aluminium, zinc, or magnesium is present. See Table R, col. 4.

(3) A *sparkling combustion of the charcoal* takes place. Probably a nitrate or a chlorate is present. See Table W.

(4) A *smell of burning sulphur* is perceived. Probably sulphur or a sulphide or sulphate is present. See Tables X and Z.

243. *To detect metals by the colors which they impart to a bead of borax-glass.*—Take a piece of platinum wire of such a thickness that three inches weigh one grain, and seal it into a glass handle as described at p. 78. Bend the end of it round into a small loop (fig. 77), which must not be larger than the transverse section of the blowpipe-flame.

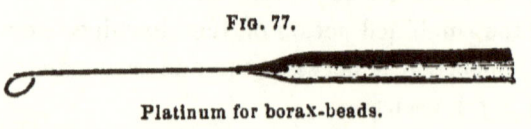

Fig. 77.
Platinum for borax-beads.

Make this loop red hot, and dip it into powdered borax, of which a considerable quantity will stick to the wire. Fuse this in the blowpipe-flame (fig. 78) to a bead which should be perfectly colorless and transparent even after cooling.

Fig. 78.

Borax-bead test.

Place one or two small particles of the substance under examination upon a piece of paper close at hand, heat the borax-bead, and touch one of the particles with it so that it may stick to the bead.

Fuse the bead at the extreme point of the outer (oxidizing) flame of the blowpipe, until the substance appears to have been dissolved by the borax, and observe the color of the bead while hot and after cooling. Should the bead be opaque, too large a quantity of the substance has been added, and the experiment must be repeated.

If the bead remains colorless, another particle of the substance must be fused with it, and so on, until a reasonable proportion has been added without producing any result.

The bead is afterwards exposed to the point of the inner (reducing) flame of the blowpipe, and the color of the glass carefully observed.

The results are compared with Table T.

To clean the wire for a new test, dip the red-hot bead into water, when it will become brittle and may be easily detached from the loop, which is then opened and the wire scraped clean with the thumb-nail.

244. *To detect metals by the color which they impart to the blowpipe-flame.*—Take a straight piece of platinum wire, as recommended at p. 78, dip it in hydrochloric acid,* and expose it repeatedly to the point of the inner (reducing) blowpipe-flame till it no longer imparts a distinct color to the outer flame. Again moisten it with hydrochloric acid, take upon it a minute quantity of the substance under examination, and again expose it to the point of the inner flame (fig. 79).

FIG. 79.

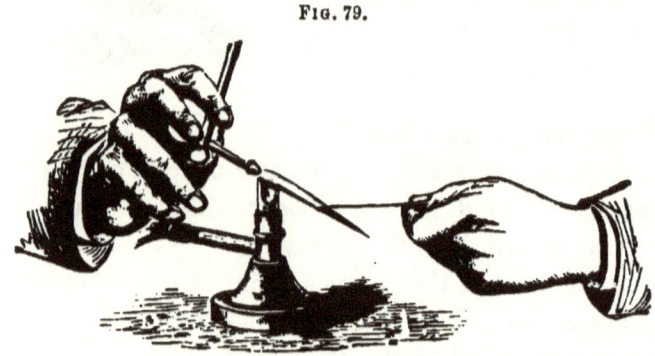

Compare the results with Table U.

Some metals (barium, for example) so fuse into the platinum that it is scarcely possible to get rid of them in order to prepare the wire for a fresh test. When this is found to be the case, the extremity of the wire must be cut off.

245. *To detect metals by the cobalt test before the blowpipe.*—The solution of nitrate of cobalt should contain about one part of the salt dissolved in ten parts of water, and is conveniently kept in a bottle provided with a perforated cork carrying a piece of glass tube long enough to reach to the bottom of the bottle and to project an inch or two above the cork (fig. 80). This tube should be partly closed at each end by directing the blowpipe-flame upon it. When the upper orifice of this tube is closed with the finger (fig. 81), and the tube withdrawn from the bottle, the cobalt solution is retained

---

* This wire should not be dipped into the bottle of hydrochloric acid, but into a small quantity of the acid in a small glass or porcelain capsule.

FIG. 80.

FIG. 81.

in it by atmospheric pressure, and may be suffered to fall, drop by drop, upon the mass under examination. One or two drops will generally suffice. An intense heat should then be applied for several seconds, the result being compared with column 4 of Table R, p. 190.

246. *To detect mercury, arsenic, and ammonia by the blowpipe.*—Dry the carbonate of soda by heating it moderately on a piece of glass or tin-plate, or on the blade of a spatula. Allow it to cool, and mix it with the powdered substance under examination, upon a piece of paper, using at least six times as much carbonate of soda as of the substance. Scrape enough powder off a piece of charcoal to impart a dark-gray color to the mixture.

FIG. 82.

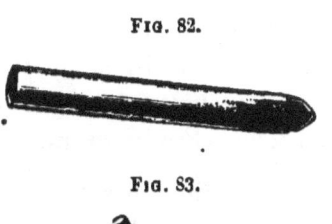

FIG. 83.

Introduce the mixture into a small dry German glass tube (p. 33), cleanse the upper part of the tube from adhering particles of substance with a match stick or a roll of paper, and rap the tube against the table so as to shake the powder into the position shown in fig. 82, leaving a clear passage for the vapor. Hold the tube in a narrow band of folded paper,* and heat it in the non-luminous part of the flame (fig. 83) so as not to smoke it. If none of the results mentioned in column 5 of Table R (page 190) are observed, direct the blowpipe-flame upon the bottom of the tube until the glass begins to fuse.

Beginners frequently fail in this test, in consequence of their allowing condensed water to trickle back upon the hot mixture, causing it to spirt up and soil the sides of the tube.

NOTES TO TABLE R.

247. Great care is necessary to avoid error in applying the blowpipe test for aluminium.

The solution of nitrate of cobalt itself leaves a blue mass of the anhydrous nitrate when evaporated to dryness by the heat of the flame; but when this blue mass is strongly heated, it is converted into the black oxide of cobalt, whereas the blue mass furnished by the combination of alumina with oxide of cobalt (aluminate of cobalt) becomes of a brighter blue color when more intensely heated.

The alkaline phosphates and borates also give a blue mass with nitrate of cobalt, but this mass assumes the condition of a fused glass, whilst the alumina blue is perfectly infusible.

To ascertain what compound of aluminium is under examination, the analyst should refer to p. 59.

248. The presence of zinc should be confirmed by observing that the infusible mass, before the addition of nitrate of cobalt, is yellow while hot, and becomes white on cooling.

To ascertain what compound of zinc is under examination, the analyst should refer to p. 61.

---

* If any moisture is seen to condense on the sides of the tube, hold the latter with its mouth somewhat inclined downwards, lest the drops should run back and crack the glass.

249. The pink color of the compound formed by magnesia with oxide of cobalt is very pale, and might be overlooked in a cursory inspection, especially if the observer be dazzled by the incandescence of the ignited mass.

To ascertain what compound of magnesium is under examination, the analyst should refer to p. 23.

250. The globules of mercury composing this sublimate are sometimes so minute that it has the appearance of a gray coating upon the tube; on rubbing it with a lucifer-match stick, the particles unite into distinct globules.

To ascertain what compound of mercury is under examination, the analyst should refer to pp. 27 and 40.

251. If arsenic be present, a garlic odor will generally be perceptible at the mouth of the tube, and should have been already noticed in column 1.

A black shining ring is sometimes formed upon the sides of the tube, in cases where no arsenic is present, in consequence of the deposition of a coating of lustrous carbon from the action of heat upon tarry matters, distilled, either from imperfectly prepared charcoal, or from organic matter in the substance analyzed.

Sulphide of mercury may also furnish a black ring if the quantity of carbonate of soda employed in the experiment is insufficient to remove the whole of the sulphur.

To ascertain that the ring does really consist of arsenic, make a deep file-mark on each side of it, and break off the glass with a jerk. Wrap the piece of glass containing the ring in a piece of stout paper, and break it, with a few blows, into fragments, but not into powder. Select those fragments which appear to be well-coated, place them in a small tube sealed at one end, and heat them gently above the point of an ordinary flame as shown in fig. 84. The arsenic will be slowly converted into vapor, combining with the oxygen of the air in the tube to form arsenious acid, which will be deposited in small brilliant crystals upon the cooler part of the tube. Under the microscope, the octahedral shape of these crystals will be distinctly seen (fig. 85).

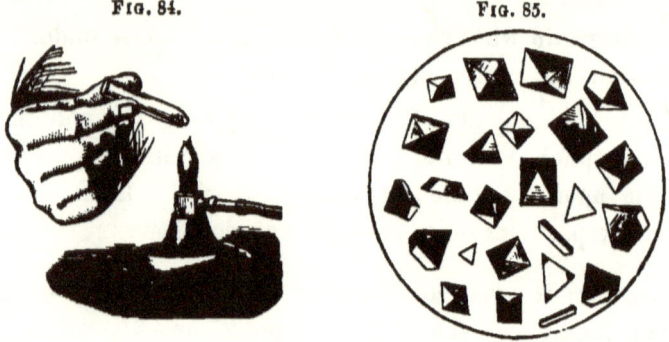

Fig. 84.   Fig. 85.

If there be more arsenic than can be oxidized by the air in the tube, a dark ring of arsenic will accompany the crystals, and may be oxidized by a second sublimation.

The smallest fragment of arsenious acid may be identified with perfect certainty in the following manner:

Draw out a piece of narrow German tube to a long point (17), and seal it in the blowpipe-flame. Drop into this point the fragment supposed to be arsenious acid. Place three or four little pieces of charcoal in the wider part of the tube (fig. 86), and heat them to redness in the blowpipe-flame. Then heat quickly the point of the tube so as to drive the vapor of arsenious acid over the red-hot charcoal, which will remove the oxygen of the arsenious acid, and a shining black ring of arsenic will be deposited upon the cooler part of the tube.

Fig. 86.

To ascertain what compound of arsenic is under examination, the analyst should refer to (35).

252. An odor of ammonia might also arise in this experiment, from the decomposition by heat of organic matters containing nitrogen; but such substances would be carbonized or blackened when heated in a small tube by themselves, and would evolve vapors having the offensive smell of singed hair, and restoring the blue color to red litmus paper.

To ascertain what compound of ammonium is under examination, the analyst should refer to (75).

TABLE S.

253. Metals reduced on Charcoal.

	Metallic Globules.	Incrustations.	Remarks.
Antimony	Very brittle	White	Metal volatilizes (254, 255).
Arsenic	None	White	Garlic fumes (256).
Bismuth	Brittle	Yellow	Metal very fusible (257).
Copper	Red, malleable	Little or none.	Difficult to fuse (258, 259, 260).
Lead	Soft, malleable	Yellow	Metal marks paper (261, 262, 263).
Silver	Malleable	Little or none.	Not oxidizable (264, 265).
Tin	Malleable	Little or none.	Easily oxidizable, very fusible (266).
Zinc	None	Yellow, hot; white when cold	Infusible mass emitting a greenish-white light (267).

NOTES TO TABLE S.

254. Antimony is generally reduced by the blowpipe in minute globules, which are not easily united into a large globule. It is easier to obtain a large globule of antimony if cyanide or ferrocyanide of potassium is employed instead of carbonate of soda.

Antimony generally gives off a white smoke, and covers the charcoal with an incrustation even beyond the limits of the oxidizing flame. The incrustation generally appears bluish-white, the black charcoal being seen through it.

The globule of antimony is extremely brittle, falling to a metallic powder when struck or pressed. The metallic appearance of the powder must be observed, or else brittle globules of slag may be mistaken for antimony.

255. The compounds of antimony most likely to be met with in blowpipe analysis, are—

Sulphide of antimony,   Tartar emetic,   Oxide of antimony.

These substances have been described at (39).

To that description it may here be added, that—

Sulphide of Antimony gives a brown mass when fused with carbonate of soda on charcoal (242).

Tartar emetic crackles (*decrepitates*) and sparkles (*scintillates*) when heated alone on charcoal, at the same time giving the violet flame of potassium.

256. The presence of arsenic should be confirmed as in column 5, Table R.

When arsenic has been discovered, (35) should be referred to, in order to ascertain whether the substance under examination is one of those described there. If it is not then identified, the blowpipe examination must be continued, with the view of discovering another metal, since arsenic is often present in considerable quantity in ores of which it does not form the most important ingredient.

257. Bismuth is often mistaken for lead by beginners, because a globule of this metal is flattened when struck or pressed, as if it were malleable; but if the flattened bead be touched with the point of the knife, it will be found to break to pieces, which is never the case with lead.

Should any doubt exist as to the presence of bismuth,

heat a portion of the substance, on charcoal, with a mixture of equal parts of sulphur and potassium iodide, when bismuth will give a bright red incrustation on the cooler part of the charcoal.

The compounds of bismuth which are most frequently met with, are described at (29).

258. Although compounds of copper are easily reduced to the metallic state before the blowpipe, a beginner generally experiences some difficulty in obtaining a globule of malleable copper. The metal first presents itself as a red-brown mass, apparently infusible; but if this be exposed to the point of the blue inner flame, which is the hottest region or *focus* of the blowpipe-flame, it fuses into a metallic globule of tough copper which requires a sharp blow to flatten it.

This experiment affords an excellent lesson in finding the focus of the flame.

259. Copper compounds containing sulphur or arsenic yield a brittle globule when reduced; to avoid this, such compounds must be well *roasted* before adding carbonate of soda, by exposing them, on charcoal, to the point of the outer (oxidizing) flame so long as they exhale any odor of sulphurous acid, or any garlic odor indicating arsenic.

260. The distinctive characters of the principal compounds of copper have been mentioned at (27).

Beside these, the blowpipe analyst is likely to meet with

Copper pyrites, composed of copper, iron, sulphur.
Gray copper ore, " { copper, iron, antimony, arsenic, and sulphur.

Copper Pyrites, $Cu_2S.Fe_2S_3$, may generally be recognized by its *yellow color* and metallic lustre; some of its varieties (*peacock ore*) exhibit beautiful rainbow tints. The presence of iron may be proved either by extracting the magnetic particles from the slag and surrounding charcoal by levigation, as directed in (241), or by fusing a particle of the ore with a bead of borax (243), in a very hot reducing (inner) flame, when the ferrous oxide will impart a bottle-green color to the glass, if the temperature be sufficiently high to fuse the copper into a minute globule which separates from the borax.

Gray Copper Ore has a *steel-gray color* and *metallic lustre.*

If it be heated alone on charcoal before the blowpipe, the sulphur and arsenic may be recognized by their characteristic odors, and the antimony by its white incrustation.

To obtain a bead of malleable copper from gray copper ore, it must be well roasted (in powder) in the outer flame until no more smell of sulphur or arsenic is perceptible, then reduced in the inner flame, with addition of carbonate of soda, and the metallic globule exposed to an intense and prolonged heat at the point of the inner flame to volatilize the antimony.

Brass (Copper and Zinc) is recognized by its giving the incrustation of oxide of zinc, which is yellow when hot, and becomes white as it cools. All the zinc may be expelled by protracted heating in the point of the inner flame, and malleable copper will be left.

Gun-metal (Copper and Tin) yields a globule which has externally the color of copper, but it is very much harder. By judicious exposure to the outer flame the bulk of the tin may be oxidized; and if a little more carbonate of soda be then added, and the reducing flame applied, a globule of nearly pure copper may be obtained.

261. Lead is very easily recognized in this experiment by the facility with which a globule of metal of considerable size may be obtained, and by the softness of the globule, which admits of being flattened out and cut with a knife. If the flattened globule be taken between the finger-nails, or upon the point of a knife, and drawn across a piece of paper, it leaves a pencil-mark.

262. The principal compounds of lead may be identified by the characters described at (14).

Galena (sulphide of lead), the principal ore of lead, when fused on charcoal with carbonate of soda, gives a brown slag containing sulphide of sodium, and if an insufficient quantity of carbonate of soda be employed, the globule of lead is often somewhat brittle from the presence of sulphur. This may be avoided by *roasting* the powdered galena in the outer flame as long as it smells of sulphur, before adding the carbonate of soda. Cyanide or ferrocyanide of potassium will furnish malleable lead with galena, without previous roasting.

Type-metal and *shrapnel bullets* are composed of lead alloyed with one-fourth of antimony, which renders the alloy hard and brittle. By exposing it for some time to the point of the reducing (inner) flame, all the antimony may be expelled in vapor, and soft malleable lead obtained.

Pewter and *Solder*, alloys of Lead and Tin, are recognized by their yielding a white infusible dross of binoxide of tin when heated in the outer blowpipe-flame.

263. *Detection of Silver in Lead by the Blowpipe. Cupellation.*—Scoop a bowl-shaped cavity about half an inch in diameter, in a piece of charcoal. Fill this with finely powdered bone-ash, which must be pressed down with the finger so as to fill the cavity with a compact mass of bone-ash, smooth and slightly hollowed on the surface. Place upon this a small piece of the lead under examination, hold the charcoal quite horizontally, and direct the extreme point of the outer (oxidizing) flame upon the metal (fig. 87). The lead immediately fuses, and the oxide which is formed is absorbed, in the liquid state, by the porous mass of bone-ash, leaving the surface of the globule covered with a very thin film of oxide which ex-

Fig. 87.

hibits rainbow tints. When this play of color on the surface of the metal ceases, and the globule is no longer diminished in size, it consists of pure silver, which is often no more than a mere speck, though easily recognized by its brilliant whiteness.

It is sometimes necessary to remove the lead from dross which has collected round it, and to transfer it to a fresh *cupel*, or bed of bone-ash.

When the lead contains *copper*, the surface of the bone-ash exhibits, after cooling, a green stain, whilst pure lead imparts a yellow color.

The presence of tin obstructs the cupellation, an infusible dross of binoxide of tin being formed upon the surface.

Silver may also be detected in bismuth by cupellation.

264. Silver is very easily reduced to the metallic state before the blowpipe, but a sharp heat is required to fuse it into a globule, which is then easily distinguished from all other metals, by its retaining its bright metallic surface when fused at the point of the outer (oxidizing) flame, and by its characteristic white color. The surface of the silver globule is generally rough and *frosted* after cooling, in consequence of the disengagement of occluded oxygen in the act of solidifying, but it acquires a strong metallic lustre when rubbed.

The compounds of silver most commonly met with have been described at (12).

When *Chloride of Silver* is fused with carbonate of soda on charcoal, the surface of the latter generally becomes coated with a white incrustation due to the condensation of some vapor of chloride of sodium.

265. *Detection of Copper in Silver.*—Cupel a small piece of the silver with a piece of pure lead, as directed at (263), when the copper will be recognized by the green stain produced on the surface of the bone-ash.

Silver containing copper is blackened when heated in the

outer blowpipe-flame, becoming covered with a film of black oxide of copper.

266. It is not easy to obtain a large globule of tin by fusion on charcoal with carbonate of soda. An infusible mass is commonly formed, which can only be reduced by repeated addition of carbonate of soda and protracted blowing. Minute globules of tin are then easily perceived in the liquid mass, but these are invisible when the slag has solidified on cooling. By levigating the mass (p. 192) large spangles of tin may be obtained.

If there be any reason for suspecting the presence of tin, it is advisable to employ coarsely-powdered cyanide of potassium instead of carbonate of soda, when a very liquid slag is obtained, in which a large globule of tin may be formed without difficulty. The cyanide of potassium, KCN, abstracts the oxygen and is converted into cyanate of potash, KCNO. Cyanide of potassium covers the charcoal with a white incrustation.

When cyanide of potassium is not at hand, powdered ferrocyanide of potassium (yellow prussiate of potash) may be substituted for it, but the globules of tin then obtained will be rather harder and less fusible, on account of the presence of a little iron. A slight yellow incrustation will also be perceived on the surface of the charcoal.

The chief compounds of tin have been described at (23) and (37). When *metallic tin* is heated on charcoal before the blowpipe, it yields a decided incrustation which is yellow while hot and white on cooling, and might lead this metal to be mistaken for zinc, but further examination prevents the error (267). This only happens when the unprotected tin is exposed directly to the strong current of the blowpipe flame; when covered with slag, tin is remarkably fixed even at a very high temperature, and hence yields little or no incrustation.

Tin-plate (sheet-iron coated with tin) is at once recognized before the blowpipe by the infusibility of the iron, the tin readily fusing and oxidizing upon its surface.

267. Zinc, being easily converted into vapor at a bright red heat, does not yield a metallic globule before the blowpipe, but the formation of the yellow incrustation which is white on cooling enables this metal to be easily detected. By moistening the incrustation with *weak* solution of nitrate of cobalt, and heating intensely, the *green* compound of oxides of cobalt and zinc may be produced.

Compounds of zinc, fused with carbonate of soda, yield eventually an infusible mass which is brilliantly *incandescent*, that is, emits a white light (with a greenish tinge) when in the flame. This mass is yellow while hot and becomes white as it cools, and if it be moisened with nitrate of cobalt (245), and again intensely heated, it becomes bright green.

The principal compounds of zinc have been described at (53). *Metallic zinc* burns explosively when heated on charcoal with the blowpipe, yielding thick white fumes of oxide of zinc.

Table T.

268. Colored Beads with Borax.

	In outer flame.	In inner flame.
Chromium...	Yellowish-green glass..	Emerald green glass (269).
Cobalt	Blue glass	Blue (270).
Copper......	Blue glass	Brown or colorless (271).
Iron	Brownish-yellow glass.	Bottle green (272).
Manganese	Purple or pink glass ...	Colorless (273).
Nickel	Brownish-yellow glass	Muddy gray (274).

NOTES TO TABLE T.

269. Should the color of the borax-bead leave any doubt as to the presence of chromium, apply the following test:—

Moisten the loop of platinum wire in the mouth, dip it into dried carbonate of soda, and fuse this into a bead, repeating the operation if necessary until the bead fills the loop. (The bead of carbonate of soda becomes opaque on cooling.) Heat the bead to redness, and take upon it a little nitrate of potash and a particle of the substance under examination. Fuse it for a few moments in the outer blowpipe-flame. If chromium be present, a bright *yellow opaque bead* will be formed, owing its color to the chromate of soda.

Should the bead have a *blue color* (green when hot), it is due to manganate of soda, and indicates the presence of manganese.

The compounds of chromium commonly met with have been described at (61).

The yellowish tinge of the chromium borax-bead in the outer flame is due to the presence of a little chromic acid, CrO_3, which becomes reduced to chromic oxide, Cr_2O_3, in the inner flame.

270. The coloring power of cobalt is very intense, and beginners often take so large a quantity of the substance upon the bead that a black opaque glass is obtained.

The presence of a little iron renders the cobalt bead green while hot.

By reduction on charcoal and levigation (241) many of the compounds of cobalt may be made to yield grains of metallic cobalt which might be mistaken for iron, but they are dissolved by a mixture of hydrochloric and nitric acids to a blue or green solution, which becomes pink when diluted.

The principal compounds of cobalt have been described at (47).

Cobalt ores generally contain arsenic, sulphur, bismuth, copper, iron and nickel, as well as cobalt.

271. The bead furnished by copper in the outer flame is green while hot, and becomes blue on cooling, but the shade of blue is very different from that given by cobalt. To render the glass colorless in the inner flame, it is necessary that the bead should not be very highly charged with copper, and should be exposed just in the hottest part, at the extremity of the inner flame, so that the reduced copper may fuse into a small bead and attach itself to the wire. If there be too much copper, or the bead be not sufficiently heated, the bead will assume an opaque brown color, due to the separation of minute particles of copper throughout the mass.

If a small particle of cyanide of potassium be taken upon the hot bead of borax which has been charged with a compound of copper in the outer flame, and the bead be then held for a second or two in the smokeless part of the ordinary flame, it at once becomes opaque brown, and if now fused in the outer blowpipe-flame, it becomes opaque blue after cooling.

The presence of copper should always be confirmed by reduction on charcoal with carbonate of soda.

The chief compounds of copper have been described at (27).

272. The colors assumed by the iron-bead in the outer and inner flames will be found to vary much with the quantity of iron present.

The brownish-yellow glass obtained in the outer flame fades very much on cooling, so that if little iron is present it often becomes colorless.

Since iron is often found in small quantity in substances of which it is not the most important ingredient, the analyst must use some judgment in arriving at a conclusion with respect to the real nature of the substance.

The principal compounds of iron are described at (45).

Sulphate of iron, when strongly heated in a small tube

closed at one end (17), evolves thick white vapors of sulphuric acid, which strongly redden blue litmus, and are accompanied by an odor of sulphurous acid. The residue left after strongly heating will exhibit, when cold, the characteristic red color of colcothar.

273. Manganese has so great coloring power that even a small quantity will often render the glass opaque. In order to obtain a colorless glass in the inner flame, the bead must be held just at the point of a good inner flame, and the decoloration of this glass affords a good test of the skill of the operator in preserving the distinction between the two flames. If the bead be carefully observed whilst in the reducing flame, it will be seen to become streaky, and at the moment of the disappearance of these streaks the bleaching will be found complete.

If a minute particle of cyanide of potassium be taken upon the bead of borax which has been charged with manganese in the outer flame, and the bead be heated for a second or two in the smokeless part of an ordinary flame, it becomes quite colorless.

The presence of manganese should be confirmed by fusing a small particle of the substance with carbonate of soda and nitrate of potash (269) in the outer flame, when an opaque blue bead (manganate of soda) will be obtained (green when hot), which becomes brown when exposed to the inner flame.

The principal compounds of manganese have been described at (55).

274. The borax beads of iron and nickel are very commonly mistaken for each other by beginners, although close observation recognizes the peculiar muddy-gray color produced by the inner flame in consequence of the reduction of a portion of the nickel.

If a minute particle of nitrate of potash be taken up with the hot bead, and the latter exposed to the outer flame, it assumes a purple tint due to the formation of borate of nickel and potassium.

The ordinary compounds of nickel have been described at (49).

Most *nickel ores* contain arsenic, sulphur, copper, iron and cobalt, as well as nickel, and since the cobalt has much more coloring power than nickel, it is rarely that the latter metal can be detected in the ore by the blowpipe.

Nickel-speiss is a compound of arsenic, sulphur, nickel, and often iron and copper, obtained in the preparation of smalt-blue from cobalt ores. It is a dark-gray or greenish-gray mass having a metallic lustre. The arsenic and sulphur can be recognized by the odor evolved when the substance is roasted on charcoal at the point of the outer flame.

Table **U.**

275. Colored Flames.

Barium................................	Green flame* (276).
Calcium (lime)	Red (277).
Copper	Bluish-green* (258).
Potassium...........................	Violet blue† (279).
Sodium...............................	Yellow (280).
Strontium...........................	Carmine (281).

* Boracic acid also tinges the flame green, though of a different shade from that produced by barium or copper. The distinctive characters of boracic acid are given at (117). Zinc gives a greenish tint to the flame, and is sometimes mistaken for barium by beginners (267).

† Arsenic gives a livid blue color to the flame, which is mistaken by beginners for that produced by potassium; a white smoke of arsenious acid issues from the arsenic flame.

NOTES TO TABLE U.

276. A beginner very often fails to obtain the green flame of barium, especially when operating upon the sulphate. To insure success, it is well to mix a little of the substance to a paste with a drop of concentrated hydrochloric acid, and to take a *very little* of this upon the extremity of the wire, which is then held in the point of a sharp blue inner flame. At first the yellow sodium flame is generally seen, but after a short time the grass-green barium flame makes its appearance.

The green flame may be even more easily obtained by fusing the sulphate of baryta with a bead composed of fluor-spar fused with about twice as much sulphate of lime.

The tint of the barium flame is very different from that given by copper, and copper should already have been excluded by the tests on charcoal, and with the borax-bead.

The commonest compounds of barium have been described at (65).

277. It is not easy to distinguish between the flames of calcium and strontium, unless they are seen side by side. By comparing the color imparted to the flame by nitrate of strontia, with that caused by the substance under examination, the danger of error is much diminished.

278. The compounds of calcium or lime ordinarily met with have been described at (69).

Sulphate of Lime may be recognized by mixing it with an equal quantity of powdered fluor-spar (fluoride of calcium), and heating on a loop of platinum wire, when it fuses easily to a clear glass, which becomes opaque on cooling.

Phosphate of Lime may be identified by moistening it with a drop of strong sulphuric acid, and exposing it, on platinum wire, to the inner blowpipe-flame, when the phosphoric acid undergoes reduction, and the phosphorus imparts a peculiar livid greenish hue to the outer flame.

If phosphate of lime be heated in a small tube (17) with metallic magnesium, and the mass, after cooling, be moistened with water, it evolves the peculiar fishy odor of phosphoretted hydrogen.

Fluoride of Calcium, when heated on a knife, or on a piece of platinum foil, generally crackles (*decrepitates*) and emits a peculiar blue light resembling a pale flame of sulphur. When mixed with an equal quantity of sulphate of lime and heated on a loop of platinum wire it fuses easily to a clear glass, which becomes opaque on cooling.

279. The violet-blue flame of potassium is easily obscured by the yellow sodium-flame. To avoid error from this cause, the flame should be examined through a square of cobalt-blue glass, through which the yellow rays of the sodium-flame are not transmitted.

The common compounds of potassium have been described at (77).

Nitrate and Chlorate of Potash would cause vivid combustion (*deflagration*) when heated on charcoal.

Bisulphate of Potash gives thick suffocating fumes of sulphuric acid when heated before the blowpipe on charcoal.

Bitartrate of Potash emits the characteristic smell of burnt sugar when heated before the blowpipe.

Ferrocyanide of Potassium (yellow prussiate of potash) evolves an odor of ammonia when heated.

280. Since almost every substance contains enough sodium to impart a yellow tinge to the flame, the analyst must hesitate before deciding that sodium is an essential constituent of the substance under examination. In order to warrant such a conclusion, the substance must impart a very strong and persistent yellow color to the flame.

It should also be carefully examined for those characters which are described at (80) as belonging to the ordinary compounds of sodium.

Chloride of Sodium usually crackles (*decrepitates*) when heated.

Nitrate of Soda causes vivid combustion (*deflagration*) when heated on charcoal.

Hyposulphite of Soda, when heated, melts easily, burns with a blue flame evolving the odor of burning sulphur, and leaves a brown residue of sulphide of sodium.

Biborate of Soda would be identified by its yielding a borax-bead on a loop of platinum wire.

Tungstate of Soda, $Na_2WO_4.2Aq.$, may be recognized by fusing it with a bead of borax, to which the tungstic acid imparts a pale-yellow color in the outer flame, becoming indigo-blue in the inner flame.

281. The precaution recommended in (277) should be taken in order to avoid mistaking calcium for strontium.

The chief compounds of strontium have been described at (67).

282. EXAMPLES FOR PRACTICE IN EXERCISE XVI. (See 10.)

Arsenious acid	Colcothar (peroxide of iron)
Chloride of ammonium	Oxide of bismuth
Binoxide of tin	Binoxide of manganese
Red lead	Oxide of cobalt
Chloride of silver	Nitrate of strontia
Oxide of antimony	Nitrate of potash
Oxide of nickel	Carbonate of soda
Chloride of barium	Lead containing silver and copper*
Carbonate of lime	
Calomel	Oxide of chromium
Oxide of zinc	Alum
Oxide of copper	Sulphate of magnesia.

* Obtained by melting a pound of lead in a crucible with a little borax, dissolving a shilling in it, casting the alloy into a thin plate, on a stone, and cutting it up into fragments.

EXERCISE XVII.—(For Examples for Practice, see 307.)

283. Detection of Non-metals or Acids by the Blowpipe.

TABLE V.

1.	2.	3.
Mix the powdered substance with about an equal quantity of BISULPHATE OF POTASH, and heat in a small tube (17). Observe whether there is any effervescence, and whether any gas having a peculiar smell or color is evolved. Compare the results with Table W. If no result is obtained, pass on to column 2.	Heat the substance by itself in a small tube (17). Observe whether it blackens, or emits any gas having a peculiar color or odor.* Test the evolved gas or vapor with BLUE AND RED LITMUS PAPER. Compare the results with Table X. If no result is obtained, pass on to column 3.	Heat the substance on charcoal by the blowpipe-flame, first alone, and afterwards with CARBONATE OF SODA. Observe the appearances presented by the mass, and any odor which may be evolved when the substance is heated alone. Compare the results with Table Y. If no result is obtained, see Table Z.

* Sulphur and phosphorus in a free state would be recognized in this experiment.

TABLE W.

284. Bisulphate of Potash Test (285).

Chlorides	evolve	Hydrochloric acid, recognized by	{ Pungent odor { Fumes at mouth of tube (286).
Nitrates	"	{ Nitric acid and { Nitric peroxide }	{ Brown vapor { Characteristic odor (287).
Carbonates	"	Carbonic acid . .	Effervescence (288).
Fluorides	"	Hydrofluoric acid	{ Extremely pungent odor { Corrosion of tube (289).
Chlorates	"	Chlorine . . .	{ Characteristic odor { Yellow color
Hypochlorites . . .	"	" . . .	Bleaches test-papers (290).
Iodides	"	Iodine	{ Violet vapors { Peculiar odor (291).
Bromides	"	Bromine . . .	{ Brown vapor { Painful odor (292).

NOTES TO TABLE **W**.

285. In performing this test, it is seldom requisite to apply the blowpipe-flame. The tube should be held, as represented in fig. 83, in the lower part of an ordinary flame.

Bisulphate of potash, when raised to a moderately high temperature, evolves vapor of sulphuric acid. The analyst should familiarize himself with the smell of the vapor obtained by heating the bisulphate of potash, before proceeding to draw a conclusion from this test.

The action of the bisulphate of potash is due to its powerfully acid character, which enables it to expel acids from their compounds, at a high temperature, much in the same way as strong sulphuric acid itself.

286. If a little black oxide of manganese or nitrate of potash be mixed with the bisulphate of potash and the suspected chloride, it will evolve chlorine itself on applying heat, which may be recognized by its odor and its power of bleaching test-papers.

The following blowpipe test is also sometimes employed for the detection of a chloride:—

Make a bead of metaphosphate of soda by fusing microcosmic salt (phosphate of soda and ammonia) in a loop of platinum wire, as in the borax-bead test;* take upon this some black oxide of copper, and fuse it in the outer flame of the blowpipe. The bead has a dark-blue color, and does not impart any blue or green color to the outer flame when held in the inner flame. If a small quantity of a chloride be taken up on the hot bead, it will color the outer flame green or blue, when held in the inner flame, in consequence of the formation of chloride of copper. Sulphate of copper may be

---

* The loop should be made double to enable the microcosmic salt to hang to it better while being fused.

substituted for the oxide, but the color is then usually limited to a bright-blue halo immediately around the bead. Bromides and iodides would produce a similar result, but they would evolve bromine and iodine, respectively, when fused with bisulphate of potash.

The description of the principal chlorides will be found on referring to the index.

287. If a little chloride of sodium (common salt) be added to the mixture of bisulphate of potash with the suspected nitrate, chlorine will be evolved, and may be recognized by its peculiar odor and power of bleaching test-papers.

If a nitrate is heated upon charcoal, it causes vivid combustion (*deflagration*).

All nitrates except those of potash, soda, and ammonia, evolve, sooner or later, brown vapor of nitric peroxide, with its peculiar odor, when heated in a tube, by themselves.

The description of the principal nitrates will be found on referring to the index.

The ordinary saltpetre of commerce contains chlorides of potassium and sodium, and therefore evolves some chlorine when heated with bisulphate of potash.

288. The carbonates are of course far more easily recognized by their effervescence when moistened with hydrochloric acid in the colored flame test (244).

289. The odor of hydrofluoric acid is far more painfully pungent than that of hydrochloric acid.

On breathing upon the mouth of the tube, a little opaque silica is deposited on the glass, from the decomposition, by moisture, of fluoride of silicon resulting from the action of the hydrofluoric acid upon the silica in the glass.

The corrosion of the tube is indicated by its peculiar greasy appearance, but can only be fully seen after the tube has been well washed and dried.

Fluor-spar (fluoride of calcium) and kryolite (fluoride of aluminium and sodium) are the only fluorides commonly met with in blowpipe analysis, and have been described at (59) and (89).

290. It must be remembered that chlorine would also be evolved from a chlo*ride* mixed with some oxidizing agent (287).

Chlorates, heated on charcoal, cause vivid combustion (*deflagration*).

The chlorates, when moderately heated by themselves, in a small tube, evolve oxygen, recognized by its power of accelerating the combustion of a spark at the end of a match when held in the mouth of the tube.

The only chlorates at all commonly met with, those of potash and baryta, have been described at pp. 82 and 70.

Hypochlorite of lime, the only hypochlorite likely to be met with, has been described at p. 74.

291. Any doubt about the presence of iodine could be set at rest by exposing to the vapor a piece of cotton or paper which has been starched; this would be colored intensely blue by the iodine, if previously moistened. Iodine itself would be recognized by heating it in a tube, when it would fuse and be entirely converted into violet vapor, condensing in black shining scales on the side of the tube. The principal iodides will be found on referring to the index.

292. If a piece of moistened starched paper or cotton be exposed to the vapor of bromine, it acquires a fine yellow color.

Table **X**.

293. Heating in small tube for Non-metals and Acids.

Acid vapors are evolved without carbonization.	Either sulphuric, hydrochloric, nitric, or oxalic acid is probably present (294).
Carbonization takes place.	Presence of some organic matter; perhaps tartaric or acetic acid (295).
Brown acid vapors	Presence of nitric acid (296).
Reddish drops of sulphur condense on the sides of the tube.	Presence of sulphur or a sulphide (297).
Violet vapors.	Presence of iodine (298).
Cyanogen is evolved, burning with a pink flame.	Presence of a cyanide (299).
Oxygen is evolved and rekindles a spark on the end of a match.	Presence of a chlorate, a nitrate, or some easily decomposed metallic oxide (300).

NOTES TO TABLE **X**.

294. The sulphates are affected by heat in very different degrees.

Sulphates of potash, soda, baryta, strontia, lime, magnesia, and lead, would not yield acid vapors in this experiment.

Bisulphates of potash and soda give strong fumes of sulphuric acid.

Sulphate of ammonia evolves sulphurous and sulphuric acids, and is entirely dissipated by heat.

Sulphate of iron evolves sulphurous and sulphuric acids, leaving a red residue of peroxide of iron.

The sulphates of alumina, zinc, manganese, nickel, cobalt, and copper require a much higher temperature to decompose them, and evolve chiefly sulphurous acid.

295. If tartaric acid be present, a peculiar smell, like that of burnt sugar, is evolved. The acetates also evolve an odor of acetone, which is rather pleasant.

The tartrates and acetates of potash, soda, baryta, strontia, and lime, when heated, leave a residue composed of charcoal mixed with a carbonate, which effervesces with acids, and in the cases of potash and soda, is very strongly alkaline.

296. The nitrates of potash and soda do not evolve acid vapors when heated. Nitrate of ammonia is extremely fusible, and is entirely dissipated by heat.

The other nitrates evolve brown acid vapors when heated. Nitrate of silver requires a high temperature for its decomposition.

297. The smell of sulphur would be perceived here.

The only common sulphides which give off sulphur when heated in a tube are iron pyrites and copper pyrites.

298. The smell of iodine is characteristic.

Indigo blue also gives off violet vapors when heated.

Iodide of lead is the only common iodide which evolves iodine vapor when heated in a tube.

299. The smell of cyanogen is very peculiar.

This gas should burn with a pink flame on approaching the mouth of the tube to a light.

Cyanides of mercury and silver are the only ordinary cyanides which evolve cyanogen when heated.

300. Chlorates yield oxygen more abundantly and at a

lower temperature than the nitrates, and both salts fuse before undergoing decomposition.

The metallic oxides would not fuse before evolving oxygen. The principal metallic oxides which give off oxygen when heated, are black oxide of manganese, red lead, oxide of mercury, oxide of silver.

Table Y.

301. Detection of Non metals and Acids on charcoal before the blowpipe.

Nitrates *Chlorates*	Deflagrate, that is, cause vivid combustion of the charcoal.
Sulphides	Evolve the odor of sulphurous acid when roasted alone in the outer flame. Yield a brown fused mass with carbonate of soda in the inner flame.
Silicates *Borates*	Yield with carbonate of soda a bead of glass which is not absorbed by the charcoal (242).

TABLE Z.
302. Special Examination for Acids difficult of detection by the Blowpipe.

Sulphuric Acid.	*Silicic Acid.*	*Boracic Acid.*	*Phosphoric Acid.*
Mix the substance with *pure* CARBONATE OF SODA, and powdered CHARCOAL, and fuse, on charcoal, in the reducing (inner) flame. Place the mass on a SILVER COIN and moisten it with HYDROCHLORIC ACID. Sulphuretted hydrogen is evolved, which blackens the coin. (303).	Make a very small bead of CARBONATE OF SODA on platinum wire (269). Take upon the hot bead successive portions of the substance and fuse in the outer flame. *Silicic Acid* should dissolve with effervescence, and eventually render the bead transparent even after cooling. (304).	Mix the substance with some FLUOR-SPAR and BISULPHATE OF POTASH and heat the mixture on a platinum loop in the inner flame. The outer flame will be tinged green. (305).	Moisten the substance with STRONG SULPHURIC ACID, and heat, on a platinum loop, in the inner flame. A livid greenish tinge is imparted to the outer flame. (306).

* Common carbonate of soda contains sulphate of soda. The bicarbonate is usually purer, and will answer the same purpose.

NOTES TO TABLE Z.

303. As this test depends upon the formation of sulphide of sodium, any compound containing sulphur would furnish the same result, so that the analyst, before concluding that the substance under examination is a sulphate, must take its general characters into consideration.

The presence of sulphide of sodium in the fused mass is detected with greater delicacy by dissolving it, on a crucible-lid or in a small capsule, in a few drops of water, and adding a little nitroprusside of sodium, $Na_2FeNO5CN$, which gives a fine purple color.

The following is also a delicate blowpipe test for sulphur: Make a small bead of pure carbonate of soda on a platinum loop (269), and take successive portions of *pure* silica upon it, until the fused bead remains transparent after cooling. The bead should now remain colorless even when heated in the reducing flame, but if a particle of a substance containing sulphur in any form be added to it, it will assume a brown yellow color in the reducing (inner) flame, due to the formation of sulphide of sodium.

In very exact analysis, a gas-flame must not be employed in testing for sulphur, since this element is always present in coal-gas, and is absorbed from the flame by the carbonate of soda. The flame of a spirit or oil lamp is free from sulphur.

304. This test requires some patience, since it is necessary to repeat the addition of silicic acid many times in order to obtain a transparent glass of silicate of soda. The glass is often colored yellow by the presence of a little sulphide of sodium (see 303).

Boracic acid would also render a bead of carbonate of soda transparent, but the following test will prevent mistakes. Make a bead with microcosmic salt (286), which is perfectly transparent. The bead thus obtained will dissolve nearly every other substance but silica, so that particles of that substance may be seen floating about undissolved in the fused bead.

305. This extremely delicate test for boracic acid depends upon the production of fluoride of boron, by the action of the hydrofluoric acid resulting from the decomposition of the fluor-spar by the bisulphate of potash. The fluoride of boron tinges the outer flame green. If much sodium be present, the green color is not easily perceived.

If a platinum wire be moistened with glycerin and a powdered borate be taken upon it, a distinct green tinge will be imparted to the flame when the glycerin is kindled. The presence of sodium does not prevent the color from being seen.

Borax (biborate of soda) would be easily recognized by its behavior when heated on a loop of platinum wire.

306. In this test, the phosphoric acid set free by the action of sulphuric acid is deoxidized by the flame, and the vapor of phosphorus which is produced imparts the peculiar livid tinge to the outer flame.

This test does not give a satisfactory result when much sodium is present.

If a compound containing phosphoric acid be heated in a tube (17) with a little metallic magnesium, and the mass, after cooling, be moistened with water, the peculiar fishy odor of phosphoretted hydrogen will be perceived.

The following is the most conclusive blowpipe test for phosphoric acid:—

Mix the compound to be tested with two or three parts of dried boracic acid, and fuse it in a cavity scooped in charcoal. Thrust into the melted mass about half an inch of very thin iron wire (of which twelve inches weigh one grain) and heat it for two or three minutes in a powerful reducing flame. If phosphoric acid be present, a hard silvery brittle metallic globule of iron containing phosphorus will be formed, which may be extracted from the fused mass by wrapping it in stout paper and striking it with a hammer upon the anvil.

307. EXAMPLES FOR PRACTICE IN EXERCISE XVII. (10).

Chloride of sodium	Nitrate of potash
Carbonate of soda	Fluor-spar
Chlorate of potash	Iodide of potassium
Bitartrate of potash	Acetate of lead
Nitrate of lead	Iron pyrites
Cyanide of mercury	Sulphide of iron
Silica (white sand)	Boracic acid
Sulphate of baryta	Phosphate of lime (bone-ash).

ALPHABETICAL LIST OF THE PRINCIPAL TESTS OR REAGENTS.

308. ACETATE OF LEAD, or PLUMBIC ACETATE $Pb2C_2H_3O_2$, may be obtained at oil and color shops as *sugar of lead* (14). One ounce is sufficient. To prepare the solution, shake half an ounce with five measured ounces of distilled or rain water, until dissolved, with the exception of a little white carbonate of lead which may be filtered off.

Scraps of lead partly covered with vinegar, in an open bottle, will give a solution of acetate of lead in the course of a few hours.

Acetate of Lead is a test for

Sulphuretted hydrogen	Black precipitate (97)
Chromic acid	Yellow precipitate (120)
Iodides	Yellow precipitate (93).

309. ACETIC ACID, $C_2H_4O_2$, is the acid of vinegar, and is sold by druggists in a diluted state (149). Four ounces will suffice. It must not give any precipitate when mixed with *excess* of ammonia (5) and sulphide of ammonium.

By boiling vinegar with powdered wood-charcoal, and filtering it, or much better, by distilling vinegar (227), acetic acid may be obtained sufficiently pure for most purposes.

310. ALCOHOL C_2H_6O, of sufficient strength for ordinary use iss old as *spirits of wine*. The *methylated spirit* is much cheaper, and answers for most of the uses of alcohol (209, 210). Four ounces (or a gill) will suffice.

311. AMMONIA, NH_3, is sold by the druggists as *liquor ammoniæ* or *hartshorn*, prepared by absorbing ammoniacal gas in water (75). Eight ounces should be provided.

It must not give any precipitate with oxalate of ammonia (indicating lime), and very little, if any, with chloride of calcium (indicating carbonate of ammonia).

Ammonia is a test for

Copper	Bright blue color (26)
Nickel	Violet blue (48)
Iron (as peroxide) . . .	Red brown precipitate (45)
Mercury (as a mercurious salt)	Gray precipitate (13).

312. Antimoniate of Potash, or **Hydropotassic Metantimoniate,** $K_2H_2Sb_2O_7$, can be obtained only from the operative chemist (39). Two drachms of the salt should be boiled with five measured ounces of distilled or rain water, in a flask, for a few minutes, and filtered. A drop of the solution stirred with a drop of carbonate of soda upon a slip of glass (fig. 27) should give a distinct precipitate upon the lines of friction.

313. Bicarbonate of Soda, or **Hydrosodic Carbonate,** $NaHCO_3$, is sold by the druggist under the name of *carbonate of soda*, in the form of powder (80). One ounce should be placed in the bottle, and shaken with eight measured ounces of cold distilled or rain water.

Bichloride of Mercury; see Perchloride of Mercury.

314. Bichloride of Platinum, or **Platinic Chloride,** $PtCl_4$ (also called *chloride* and *perchloride of platinum* and *muriate of platina*), may be obtained from the operative chemist, generally in the form of a solution in water. It is expensive. Half an ounce will suffice. It should give a well marked precipitate when stirred upon a slip of glass with a drop of a moderately strong solution of nitrate of potash (76).

To prepare bichloride of platinum, dissolve 10 grains of scrap platinum (or old platinum foil) in a flask, by gently heating it with two measured drachms of strong hydrochloric acid, and half a drachm of strong nitric acid. Pour the solution into a dish, and evaporate it at a gentle heat (fig. 30) till it becomes a syrup which solidifies on cooling. Dissolve this in an ounce and a half of distilled or rain water.

315. Bichromate of Potash, or **Potassic Dichromate,** $K_2O.2CrO_3$, is sold by the operative chemist in crystals (61). Half an ounce of the crystals may be dissolved in eight measured ounces of hot distilled water.

316. BISULPHATE OF POTASH, or HYDROPOTASSIC SULPHATE, $KHSO_4$, is sold in a crystalline form by the operative chemist (77). It is only required for blowpipe analysis. One ounce will suffice. It is generally made in the laboratory from the residue left in preparing nitric acid.

317. BONE-ASH is made by burning bones white in a clear fire, and grinding the ash to a fine powder (56). It is sold by the operative chemist, and is only required for blowpipe analysis.

318. BORACIC ACID, H_3BO_3, is generally sold by the operative chemist in the crystallized form (117). It is only used in blowpipe analysis. Half an ounce of the crystals should be dried in an evaporating dish placed in an oven or over a low flame, and reduced to powder.

319. BORAX, $Na_2O.2B_2O_3$, is sold at the oil-shops in crystals (80). An ounce of the crystals should be heated in an evaporating dish, over a moderate flame, till the spongy mass ceases to swell up and evolve steam, when it should be reduced to powder.

320. CARBONATE OF AMMONIA, or AMMONIC SESQUICARBONATE, $2(NH_4)_2O.3CO_2$, is sold, under the former name, by the druggist, in white lumps (75). Two ounces of the salt should be reduced to powder, and shaken, in the bottle, with six measured ounces of cold distilled or rain water.

321. CARBONATE OF SODA, or SODIC CARBONATE, Na_2CO_3, is best obtained, for use in analysis, from the bicarbonate of soda commonly sold as a white powder under the name of Carbonate of Soda (80). This should be heated in an evaporating dish over a Bunsen burner (fig. 35) for a quarter of an hour, so as to drive off half its carbonic acid. After cooling, one ounce of the carbonate may be dissolved in six measured ounces of distilled or rain water. The solution, when acidulated with dilute nitric acid, should give no precipitate with nitrate of baryta (indicating sulphate of soda), or nitrate of silver (indicating chloride of sodium).

322. Chloride of Ammonium, or Ammonic Chloride, NH_4Cl, is sold under the names of *sal-ammoniac* and *muriate of ammonia*, by the druggist. The crystallized salt is purer than the lumps obtained by sublimation, which are colored brown by iron (75).

Dissolve one ounce of the salt in eight measured ounces of warm distilled or rain water.

Chloride of Ammonium is a test for
Platinum . . . . . . Yellow granular precipitate.

323. Chloride of Barium or Baric Chloride, $BaCl_2$, is sold by the operative chemist, as *muriate of barytes*, in crystals (65). The solution is made by dissolving half an ounce of the salt in five measured ounces of distilled or rain water.

324. Chloride of Calcium, or Calcic Chloride, $CaCl_2$, may be obtained from the operative chemist either in crystals, or more commonly in white lumps of dried chloride of calcium (69). One ounce may be dissolved in four measured ounces of distilled or rain water for use as a test.

Chloride of calcium is easily made by adding chalk, or whitening, or marble, to dilute hydrochloric acid, as long as it is dissolved, boiling the solution till it no longer reddens blue litmus paper, and filtering.

325. Chloride of Lime may be purchased at the oil-shops. If good, it is a dry powder which has a strong smell like chlorine, and gives off much chlorine when mixed with dilute sulphuric acid. To make the solution for testing, one ounce of the chloride is rubbed down in a mortar with eight ounces of water, and filtered.

326. Chlorine Water is made by passing chlorine gas into water. A small quantity may be quickly made by pouring the gas into a test-tube half full of water, and shaking violently after closing the tube with the thumb. By pouring the gas in three or four times, strong chlorine water may be made in two or three minutes. The gas is prepared by gently warming black oxide of manganese with strong hydrochloric acid, in a test-tube or flask.

326a. COPPER. Thin sheet copper in pieces about half an inch long and an eighth wide; or pieces of copper wire may be used; it may be brightened by immersion in nitric acid and washing.

Copper (when boiled in a solution acidified with hydrochloric acid)
is a test for

Mercury	Bright silvery coating	(30)
Arsenic	Dark gray "	(33)
Antimony	Purplish "	(32)
Bismuth	Light gray "	(28)

327. CYANIDE OF POTASSIUM, or POTASSIC CYANIDE, KCN, is sold by druggists in white lumps (77). These should be coarsely powdered in a mortar. Only small quantities should be kept in powder in a well-closed bottle. It is very poisonous.

Two ounces of cyanide of potassium will suffice for some time.

328. ETHER, $C_4H_{10}O$, may be obtained from the druggist, and must be kept in a stoppered bottle (221). One ounce will suffice. Methylated ether is less expensive than pure ether, and should be employed when large quantities are required.

329. FERRIDCYANIDE OF POTASSIUM, or POTASSIC FERRIDCYANIDE, K_3FeCy_6, is sold by the operative chemist in crystals, and is commonly called *red prussiate of potash* or *ferricyanide* of potassium (77). Half an ounce of the salt may be dissolved in five measured ounces of distilled or rain water.

Ferridcyanide of Potassium is a test for

Iron as a ferrous compound or protosalt . . Blue precipitate (45).

330. FERROCYANIDE OF POTASSIUM, or POTASSIC FERROCYANIDE, K_4FeCy_6, is often obtainable from the oil and color shops as *yellow prussiate of potash*, in crystals (77). For testing, half an ounce of the salt may be dissolved in five measured ounces of distilled or rain water.

Ferrocyanide of Potassium is a test for

| Iron | . | . | . | . | Blue precipitate (45). |
| Copper | . | . | . | . | Brown-red precipitate (27). |

331. HYDROCHLORIC ACID, HCl, may be obtained from the druggist as a solution of hydrochloric acid gas in water (105). Its common name is *muriatic acid*. It should be colorless, and strong enough to emit fumes in moderately damp air. Eight ounces of this acid should be provided.

Dilute hydrochloric acid is made by mixing the strong acid with twice its volume of distilled or rain water.

The acid must give no precipitate with hydrosulphuric acid (indicating arsenic, chlorine, sulphurous acid, perchloride of iron) or with *excess* of ammonia (5) and sulphide of ammonium (indicating iron), or with much water and chloride of barium (indicating sulphuric acid).

332. HYDROFLUOSILICIC ACID, or SILICOFLUORIC ACID, H_2SiF_6, is sold by the operative chemist in a diluted state. One ounce will suffice. It must not give any precipitate with solution of nitrate of strontia (indicating sulphuric acid) even on stirring.

333. HYDROSULPHATE OF AMMONIA, or SULPHIDE OF AMMONIUM, $(NH_4)_2S$, may be obtained in solution from the operative chemist (75). It must not give any precipitate with sulphate of magnesia (indicating free ammonia). Eight ounces of the solution should be provided, and must be kept in a well-closed bottle.

Hydrosulphate of ammonia is easily prepared by passing hydrosulphuric acid gas (334) into solution of ammonia until it no longer gives a precipitate with sulphate of magnesia.

334. HYDROSULPHURIC ACID, H_2S, is prepared by passing sulphuretted hydrogen gas into water until the stopper of the bottle is forced up when the liquid is shaken.

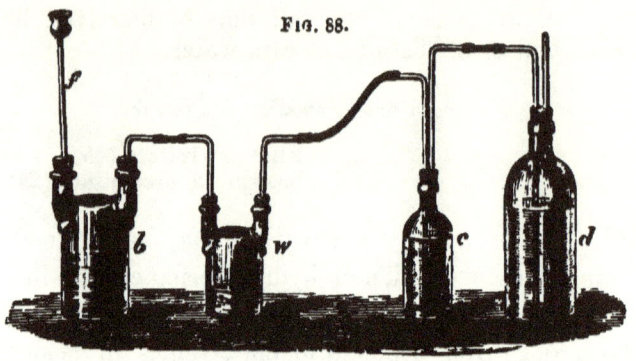

Fig. 88.

Fragments of sulphide of iron (procurable from the operative chemist) are placed in a half-pint bottle (*b*, fig. 88), which is half filled with water. Strong sulphuric acid is poured down the funnel tube (*f*) in a few drops at a time, until the bubbles of gas pass freely through the wash-bottle (*w*) which contains a little water. The gas is then passed into the bottle (*c*), which contains the distilled water to be charged with gas. Any gas which is not absorbed by the water passes into the bottle (*d*) containing solution of ammonia. This retains all the gas, and prevents it from contaminating the air. When this ammonia has become saturated with the gas, so that it no longer precipitates sulphate of magnesia, it may be used as hydrosulphate of ammonia, but it will be long before this point is reached.

About ten minutes are required to saturate the water in the bottle (*c*) with a brisk current of gas.

The parts of this apparatus are connected by vulcanized India-rubber tubing, which may be obtained from the operative chemist.

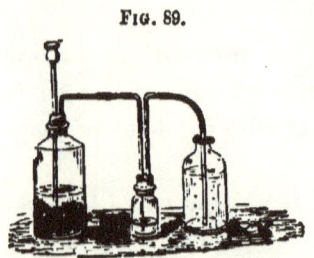

Fig. 89.

Hydrosulphuric acid apparatus.

Fig. 89 shows a simpler apparatus for preparing the solution of hydrosulphuric acid, but since there is no provision for absorbing the excess of gas, it can only be used in a fume-closet with a

Hydrosulphuric Acid is a test for

Cadmium . . .	Bright yellow precipitate in ammoniacal solution
Antimony . . .	Red precipitate in acid solution (39)
Zinc	White precipitate in ammoniacal solution (53).

335. IODIDE OF POTASSIUM, or POTASSIC IODIDE, KI, is sold in crystals by the druggist (77). It is sometimes called *hydriodate of potash*. Two drachms of the salt may be dissolved in five measured ounces of distilled or rain water.

Iodide of Potassium is a test for

Lead	Bright yellow precipitate, dissolved by boiling water (14).
Silver	Pale yellow precipitate, whitened by ammonia (12).
Mercury as a mercuric salt	Scarlet precipitate, easily dissolved by excess (31).

336. IODINE is procurable in the solid form from the druggist (93). Five grains may be shaken with an ounce of water to prepare *iodine-water*.

Iodine-water is a test for

Starch	Blue color.

337. LIME-WATER, CaH_2O_2, is sold by the druggist. It should turn reddened litmus distinctly blue. It is prepared by shaking freshly slaked lime with distilled or rain water, allowing the solution to settle in a well-closed bottle, and pouring off the clear liquid.

338. LITMUS PAPER, both red and blue, may be procured from the operative chemist.

The coloring matter is made from a species of lichen, the *Rocella tinctoria*, and is sold in blue cakes made up with chalk or plaster of Paris.

To make litmus paper, heat half an ounce of cake litmus with three measured ounces of distilled or rain water for half an hour, and filter the solution. To one-half of it, add a little *extremely* dilute sulphuric acid, on the end of a glass rod, until it acquires a faint reddish tinge (by daylight).

Add the other half of the solution, and immerse, in the blue liquid, strips of white filter-paper, or better, of unsized drawing-paper. Hang them on a string to dry, out of the reach of acid fumes. Redden the remaining liquid faintly by stirring with the rod dipped in dilute sulphuric acid, and dye some more paper with the solution, to obtain red litmus paper.

These papers fade when exposed to a strong light.

339. MICROCOSMIC SALT, $NaNH_4HPO_4$, is used only in blowpipe analysis, and may be obtained from the operative chemist. A very small quantity is used.

340. MOLYBDATE OF AMMONIA, NH_4HMoO_4, is sold by the operative chemist in crystals. It is an expensive salt. Twenty grains may be dissolved in an ounce of distilled water for testing.

Molybdate of Ammonia is a test for

Phosphoric acid . . . { Bright yellow precipitate in a nitric solution, on warming (111).

341. NITRATE OF BARYTA, or BARIC NITRATE, $Ba2NO_3$, may be obtained in crystals from the operative chemist (65). Half an ounce of the salt may be dissolved in five measured ounces of distilled or rain water.

Nitrate of Baryta is a test for

Sulphuric acid . . . . { Milky white precipitate insoluble in dilute nitric acid (102).

342. NITRATE OF COBALT, or COBALTOUS NITRATE, $Co2NO_3$, is sold, commonly in solution, by the operative chemist (47). It is required for blowpipe analysis only. A very small quantity will suffice.

343. NITRATE OF POTASH, or POTASSIC NITRATE, KNO_3, is sold at the oil-shops as *saltpetre* (77). It should be kept, for use in analysis, in powder.

344. NITRATE OF SILVER, or ARGENTIC NITRATE, $AgNO_3$, may be obtained from the druggist, in crystals, in fused sticks (*lunar caustic*) or in solution (12). Two drachms of the crystals may be dissolved in five measured ounces of distilled water.

Nitrate of Silver is a test for

Hydrochloric acid or chlorides	White precipitate insoluble in boiling nitric acid; soluble in ammonia (105).
Hydriodic acid or iodides .	Yellow precipitate insoluble in nitric acid. Whitened, but not dissolved by ammonia (106).
Hyposulphites	White precipitate, passing rapidly through various shades of yellow, red, and brown, finally becoming black sulphide of silver.

345. Nitric Acid, HNO_3, may be obtained from the druggist (109). Eight ounces should be provided, and must be kept in a stoppered bottle. It should be nearly colorless, and should fume moderately when exposed to air. After dilution with much water, it should not give any precipitate with nitrate of silver (indicating chlorine) or with nitrate of baryta (indicating sulphuric acid). *Excess* of ammonia (5) followed by sulphide of ammonium should not produce any change in it.

Dilute nitric acid may be made by mixing the strong acid with twice its volume of distilled or rain water.

346. Nitroprusside of Sodium, $Na_2FeNOCy_5$, may be procured in red crystals from the operative chemist. It is an expensive salt, very rarely required. A solution of it may be obtained by boiling ferrocyanide of potassium with diluted nitric acid till it ceases to give a blue precipitate with sulphate of iron, then adding an excess of carbonate of soda, boiling, and filtering. The solution should give a fine purple blue color with sulphide of ammonium.

347. Oxalate of Ammonia, or **Ammonic Oxalate,** $(NH_4)_2C_2O_4$, is sold by the operative chemist in crystals (75). Two drachms of the salt should be dissolved in six measured ounces of distilled or rain water.

348. Perchloride of Iron, or **Ferric Chloride,**

Fe_2Cl_6, is also called *sesquichloride of iron* and *muriate of iron*. It is obtained from the operative chemist in the state of solution in water (45).

It may be prepared by dissolving iron (wire, nails, or filings) with the aid of heat, in a mixture of dilute hydrochloric acid with one-fourth of dilute nitric acid, evaporating at a moderate heat till the liquid is syrupy, and diluting with water.

Perchloride of Iron is a test for

Ferrocyanides	Dark-blue precipitate (77)
Sulphocyanides	Blood-red solution bleached by perchloride of mercury
Tannic acid ⎫ . . . . Gallic acid ⎭	Inky-black precipitate or color (152, 153)
Hyposulphites	Red solution, soon becoming colorless.

349. PERCHLORIDE OF MERCURY, or MERCURIC CHLORIDE, $HgCl_2$, is sold by the druggist in the solid form, as *bichloride of mercury* or *corrosive sublimate* (31).

Two drachms of the salt may be dissolved in four measured ounces of water.

Perchloride of Mercury is a test for

Tin as a stannous compound or protosalt	White or gray precipitate (23)
Iodides	Scarlet precipitate (93).

349a. PERMANGANATE OF POTASH, $K_2Mn_2O_8$, is sold in solution as *Condy's disinfecting fluid*. The crystallized salt may be obtained from the operative chemist. Twenty grains of the salt may be dissolved in four ounces of water.

350. PHOSPHATE OF SODA, or HYDRO-DISODIC PHOSPHATE, Na_2HPO_4, may be obtained, in crystals, from the druggist (80). Half an ounce of the salt may be dissolved in ten measured ounces of distilled or rain water.

351. POTASH, or POTASSIC HYDRATE, KHO, is sold by the druggist both in solution as *liquor potassæ*, and in the solid form as *potassa fusa* in sticks (77). About eight ounces of the solution should be provided. One ounce of the solid potash may be dissolved in eight measured ounces of distilled or rain water.

The solution of potash should not effervesce strongly on adding an excess of hydrochloric acid (indicating carbonic acid).

The acidified solution should give but a slight flocculent precipitate (alumina) on adding ammonia in excess. Hydrosulphate of ammonia added to this solution should give little, if any, dark tinge (indicating iron, lead, or copper).

Potash is a test for

Mercury, as a mercuric salt .	Yellow precipitate (31)
Copper	Blue precipitate, becoming black when boiled with excess of potash (27)
Iron, as a ferric compound, or persalt	Red-brown precipitate (45)
Iron, as a ferrous compound, or protosalt	Dark green precipitate, becoming brown when exposed to air (45)
Manganese	White precipitate, becoming brown when shaken with air (55)
Nickel	Pale green precipitate, insoluble in excess (49)
Cobalt	Blue precipitate (47)
Chromium	Green precipitate, soluble in excess to a green solution, precipitated by boiling (61)
Ammonia	Odor of ammonia on boiling (75)

352. PROTOCHLORIDE OF TIN, or STANNOUS CHLORIDE, $SnCl_2$, should be prepared by boiling two drachms of tin with half a measured ounce of strong hydrochloric acid.

The solution may be diluted with two ounces of water, and some metallic tin should be kept in it, to prevent it from becoming converted into stannic chloride by the combination of a portion of its tin with atmospheric oxygen.

Tin is sold by the operative chemist in the granulated state, which is easily dissolved by hydrochloric acid.

Tin-foil often contains lead. If this metal be present in the solution of protochloride of tin, it will give a precipitate with dilute sulphuric acid.

Protochloride of Tin is a test for

Gold	Purple-brown precipitate
Platinum	Bright red solution
Mercury	Gray precipitate (31).

353. STARCH, $C_6H_{10}O_5$.—A drachm of white starch is rubbed down in a mortar with a measured ounce of cold water. The mixture is poured by degrees into five ounces of boiling water in a dish.

Starch is a test for

Free iodine . . . . Blue color (93).

354. SULPHATE OF COPPER, or CUPRIC SULPHATE, $CuSO_4$, is sold at the oil-shops as *blue copperas* or *blue vitriol*, in crystals (27). Half an ounce of the salt may be dissolved in five ounces of water.

355. SULPHATE OF IRON, or FERROUS SULPHATE, $FeSO_4$, also called *protosulphate of iron*, is sold at the oil-shops in crystals under the name of *copperas* or *green vitriol*, (45). One ounce of the salt may be dissolved in four measured ounces of cold distilled or rain water.

356. SULPHATE OF LIME, or CALCIC SULPHATE, $CaSO_4$, is sold at the oil-shops as *plaster of Paris* (69). About a drachm of the powder may be heated with half a pint of water for some minutes, with occasional stirring, allowed to stand till cold, and filtered off.

357. SULPHATE OF MAGNESIA, or MAGNESIC SULPHATE, $MgSO_4$, is obtained from the druggist as *Epsom salts* (9). One ounce of the salt may be dissolved in five or six measured ounces of water.

358. SULPHATE OF MANGANESE, or MANGANOUS SULPHATE, $MnSO_4$, may be obtained in crystals from the operative chemist (55). One drachm of the salt may be dissolved in two ounces of water.

Sulphate of Manganese is a test for

Hypochlorites . . . . . Dark-brown precipitate (91).

SULPHIDE OF AMMONIUM; see HYDROSULPHATE OF AMMONIA.

359. SULPHURIC ACID, H_2SO_4, is sold at the oil-shops as *oil of vitriol*, which is often of a dark brown color, and unfit for use in analysis. By carefully boiling it in a flask it may be rendered nearly colorless, the organic matter to which the brown color is due being oxidized by one portion of the acid. The flask must be allowed to cool before attempting to pour out the acid.

It is advisable, if possible, to obtain the nearly colorless oil of vitriol from the druggist. It must be kept in a stoppered bottle.

Dilute Sulphuric Acid is made by pouring one measured ounce of oil of vitriol, slowly, with continual stirring, into four measured ounces of water in an evaporating basin, beaker or flask (*not in a measure or bottle*). The mixture is allowed to cool, and the clear diluted acid poured off from the sulphate of lead which is usually deposited from the common acid.

Strong Sulphuric Acid is a test for

Chlorates	Red color, and evolution of yellow explosive gas (90)
Iodine	Violet vapors on heating (93).

360. TARTARIC ACID, $C_4H_6O_6$, is sold in crystals by the druggist (158). One ounce may be dissolved in four measured ounces of water.

361. TURMERIC PAPER may be obtained from the operative chemist. Its yellow color soon fades in a strong light.

To prepare turmeric paper, gently heat one drachm of powdered turmeric (the root of *Curcuma longa*) in six measured drachms of methylated spirit, till the solution has a bright yellow color. Filter it, and dip white filter-paper or unsized drawing-paper into it. Hang the paper on a string to dry.

362. WATER.—If it can be procured, distilled water only should be employed in analytical operations. It should be tested in the following manner, each test being performed upon a separate portion :—

1. Evaporate a few drops upon a slip of glass; scarcely a trace of solid matter should remain.
2. Add Nitrate of Silver; no turbidity (indicating chlorides or hydrochloric acid) should be produced.
3. Add Chloride of Barium; there should be no turbidity (indicating sulphates).
4. Add Oxalate of Ammonia; there should be no turbidity (indicating lime).
5. Add Hydrosulphuric Acid; there should be no dark tinge (indicating lead or copper).

Rain-water, after being filtered, may often be substituted for distilled water.

Should neither distilled nor rain water be procurable, the analyst should test the common water as directed above, and make a careful note of the results, to be subsequently employed in correcting his analysis.

Common water may often be in great measure deprived of the lime and magnesia which it contains, by boiling it gently in a kettle for half an hour, allowing it to cool, and filtering from the deposited carbonates of lime and magnesia.

Any of the arrangements represented at pp. 168–171 may be employed for distilling water.

363. Zinc may be procured at the ironmonger's. Thin sheet zinc cut up into strips will be found very convenient.

The operative chemist furnishes *granulated zinc*, made by melting the metal in an iron ladle or earthern crucible, and pouring it, in a thin stream, into a pail of water, from a height of eight or ten feet.

Apparatus required for the Qualitative Analysis of Single Substances.

364. The following list includes the principal articles which are necessary for the identification of single substances.

They may be provided for ten or twelve shillings.

12 test tubes	Narrow hard glass tubing
Rack with draining pegs	Blowpipe
Tube-cleaner	Triangular file
Spirit lamp or gas burner	Platinum wire and foil
3 funnels : 2 oz., 1 oz., $\frac{1}{2}$ oz.	4 oz. evaporating dish
Filter-paper	Half-pint Wedgwood mortar
Slips of window-glass	Wire triangle
Glass-rod	Wire tripod or retort-stand.

APPENDIX.

Qualitative Analysis of Gunpowder.
Boil with **Water** and Filter.

Undissolved	Dissolved
Charcoal, Sulphur.	Nitre (or *Nitrate of Potash*) and soluble impurities.

Undissolved	Nitre or saltpetre.			Impurities.	
	1.	2.	3.	4.	5.
	Test for *Potash* with TARTARIC ACID (76), and in a fresh portion with CHLORIDE OF PLATINUM (76).	Test for *Nitric Acid* with STRONG SULPHURIC ACID and COPPER (108).	Test for *Chlorides* with NITRATE OF SILVER. White Precipitate insoluble in NITRIC ACID.	Test for *Sulphates* with CHLORIDE OF BARIUM. White Precipitate insoluble in NITRIC ACID (dil.)	Test for *Lime* with OXALATE OF AMMONIA. White Precipitate.

1. Wash (16) till a drop of the washings leaves no residue when evaporated on glass (Fig. 5), and dry upon a hot brick.

Heat a small portion of the residue in a small tube (17). *Sulphur* sublimes in yellow or brown drops. *Charcoal* remains.

2. Heat the greater part of the residue on platinum foil. *Sulphur* burns with a blue flame and sulphurous odor. *Charcoal* glows and burns slowly away. Any residue after all the black charcoal has disappeared, consists of *Ash* and incombustible impurities.

INDEX.

ACETATE of ammonia prepared, 64
  of copper, 37
  of lead, $Pb(C_2H_3O_2)_2$, 29
  for testing, 228
Acetates, action of heat on, 133
  common, 133
Acetic acid, $HC_2H_3O_2$, 133
  confirmed, 133
  detected, 129
  for testing, 228
  identified, 133
  ether, 133
Acetone, 132
  identified, 163
Acid reaction, 34
  vapors, 221
Acids detected, 91
Acroleine, 156
Agate mortar, 119
Albumen, 154
  soluble, identified, 154
Alcohol, extracted, 167
  for testing, 228
  identified, 162
  separated from water, 167
Aldehyde identified, 163
Alkali waste, 100
Alkaline reaction, 34
Alkaloids, general test for, 179
  identified, 144
Alum, 59
  chrome, 67
  concentrated, 59
Alumina, acetate, 60
  hydrate, 58
  phosphate, 109
  silicate, 59, 127
  sulphate, 59
Aluminate of soda, 60
Aluminium and sodium, fluoride, 97

Aluminium blowpipe test, 190
  common compounds of, 59
  confirmed, 58
  detected, 51
  identified, 58
Ammonia, NH_3, 79
  acetate, 134
    prepared, 64
  alum, 59
  carbonate $(NH_4)_2CO_3$, 80
  common compounds of, 79
  detection, 77
  for testing, 228
  hydrochlorate, NH_4Cl, 79
  hydrosulphate (NH_4HS), 80
  identification, 79
  in excess, 21
  molybdate, 236
  muriate, 79
  nitrate, 80
  oxalate $(NH_4)_2C_2O_4$, 80
  sesquicarbonate, 80
  solution, 79
  sulphate, 80
  test for, 239
  urate, 142
Ammonic carbonate, $(NH_4)_2CO_3$, 80
  chloride NH_4Cl, 79
  oxalate, $(NH_4)_2C_2O_4$, 80
  sesquicarbonate, 80
  sulphide, $(NH_4)_2S$, 80
Ammonio-hydric sulphide, 233
Ammonium, 79
  blowpipe test, 190, 201
  chloride, NH_4Cl, 79
  common compounds of, 79
  sulphide $(NH_4)_2S$, 80
Amorphous phosphorus, 108
Analysis by blowpipe, 190
  liquid tests, 17
Anglesite (sulphate of lead), 30

Aniline identified, 164
  oxalate, 164
  purple, 164
  red, 164
Animal charcoal, 123
Antimoniate of potash, $KSbO_3$, 50
Antimonic acid, 126
Antimony and arsenic distinguished, 44
  blowpipe test, 201
  chloride, 50
  common compounds of, 49
  confirmed, 48
  crude, 49
  detected, 32
  flowers of, 49
  gray ore of, 49
  identified, 48
  in insoluble substances, 126
  oxide, 49
  potassio-tartrate, 49
  sulphide, 49
  sulphuret, 49
    blowpipe test, 202
  terchloride, 50
  teroxide, 49
  tersulphide, 49
  test for, 235
  vermilion, 49
Apparatus, 243
Aqua fortis, 106
Argand burner, 94
Argentic nitrate, 236
Argol, 84
Arseniate of soda, 46
Arsenic acid, 46
  and antimony distinguished, 44
  bisulphide, 46
  blowpipe test, 190, 199
  common compounds of, 45
  confirmed, 42
  detected, 32
  expelled by roasting, 203
  identified, 45
  in ores, 202
  iodide, 46
  sulphide, 46
  tersulphide, 46
  white, 45
Arsenietted hydrogen, 44
Arsenious acid, 45
  crystals, 42

Arsenious acid detected, 32
  identified, 45
  reduced, 200
Arsenite of copper, 37, 45
Astringent substances, 186
Aurum musivum, 47

BANDAGE, caoutchouc, 168
  Baric chloride, $BaCl_2$, 231
  nitrate, $Ba(NO_3)_2$, 236
Barium, 69
  blowpipe test, 212
  carbonate, 69
  chlorate, 70
  chloride, $BaCl_2$, 70
  chromate, 70
  common compounds of, 69
  confirmed, 69
  detected, 68
  nitrate, $Ba(NO_3)_2$, 70
  sulphate, 69, 125
Baryta, 70
  blowpipe test, 212
  carbonate, 69
  chlorate, 70
  chromate, 70
  confirmed, 69
  detected, 68
  hydrate, 70
  nitrate, $Ba(NO_3)_2$, 70
  oxalate, 111
  sulphate, 69, 125
    blowpipe test, 212
Barytes, 125
  muriate, 70
Bases, organic, identified, 144
Bending tubes, 172
Benzoic acid, detected, 130
  identified, 136
Benzine, 166
Benzole, 166
Benzonitrile, 136
Biborate of soda, $Na_2O, 2B_2O_3$, 89
Bicarbonate of soda, $NaHCO_3$, 87
  for testing, 229
Bichloride of mercury for testing, $HgCl_2$, 229
  of platinum for testing, $PtCl_4$, 229
Bichromate of potash, 67.
  for testing, 229
Bismuth, blowpipe test, 202
  citrate, 140

INDEX. 247

Bismuth, common compounds of, 39
    confirmed, 38
    detected, 32
    identified, 39
    nitrate, 39
    oxide, 39
    oxychloride, 39
    tested for silver, 206
    trisnitrate, 39
Bisulphate of potash for testing, $KHSO_4$, 230
    in blowpipe analysis, 218
    test, 217
Bisulphide of carbon identified, 167
Bitter almond oil, 166
    almonds, odor of, 137, 166
Black borax bead, 209
    jack, 61
    lead, 123
    oxide of copper, 38
        of iron, 55
        of manganese, 63
    sulphide of antimony, 49
Bleaching powder, 74
Blende, 61
Blowpipe analysis, 190
    borax bead test, 194
    cobalt test, 196
    colored flame test, 195
    detection of acids, 216
    for non-metals, 216
    for reduction on charcoal, 191
    detection of ammonium, arsenic, mercury, 197
    gas, 173
    reduction of metals by, 191
    spirit, 173
    test in small tubes, 221
Blue carbonate of copper, 37
    crystals, 37
    flame, 81
    fused mass, 124
    indigo, 149
    iodized starch, 155
    litmus paper, 235
    mass, with nitrate of cobalt, 198
    Prussian, 56, 111
    smalt, 57

Blue solution, 37, 57
    stone, 37
    Turnbull's, 52
    ultramarine, 100
    vitriol, 37
Boiling, 19
Bone-ash, 64
    for cupellation, 230
Bone-black, 123
Boracic acid, blowpipe test, 226
    confirmed, 112
    crystals, 113
    detected, 93
    for testing, 230
    identified, 113
    Tuscan, 113
Borate of lime, 113
Borates, blowpipe test, 226
    detected, 93
    on charcoal, 193
Borax, $Na_2O.2B_2O_3$, 89
    bead test, 194
    beads, 208
    blowpipe test, 226
    for testing, 230
Boron, fluoride, 226
Boronatrocalcite, 113
Brass, 204
Brick, 127
Brimstone, 122
British gum, 155
Bromides, blowpipe test, 220
Bromine detected, 96, 103
Bronze powder, 47
Brown acid vapors, 221
    caramel, 149
    chromate of lead, 30
    color with potash, 154
    ferrocyanide of copper, 111
    hæmatite, 54
    oxide of lead, 29
    residue with nitric acid, 29
    vapors, 103
Brucine detected, 144
    identified, 146
Brunswick green, 38
Bullets, shrapnel, 205
Bunsen's burner, 75, 95
Burnett's disinfecting fluid, 62
Burnt sugar odor, 137
Butyric acid identified, 163
    ether, 164

CADMIUM, test for, 235
  Caffeine detected, 145
Caffeine identified, 145
Calamine, 61
  electric, 128
Calcic chloride, CaCl$_2$, 74
  sulphate, CaSO$_4$, 73
Calcined magnesia, 24
Calcium, 71
  blowpipe test, 212
  carbonate, 72
  chloride, CaCl$_2$, 74
  common compounds of, 72
  detected, 68
  fluoride, 65
    blowpipe test, 214, 219
  oxalate, 74
  phosphate, 64
    blowpipe test, 213
  sulphate, CaSO$_4$, 73
    blowpipe test, 213
  sulphide, 100
Calomel, 27
Cane sugar identified, 152
Caoutchouc bandage, 168
Caramel identified, 149
Carbazotic acid identified, 149
Carbolic acid, 164
  identified, 150
Carbon, 122
  identified, 116
Carbonate of ammonia, (NH$_4$)$_2$CO$_3$, 80
  for testing, 230
  precipitate, 68
  copper, 37
  lead, 29
  lime, 72
  magnesia, 24
  soda, Na$_2$CO$_3$, 87
    bead, 209
    for testing, 230
Carbonates, blowpipe test, 217
  common, 99
Carbonic acid, detected, 91
  identified, 99
  oxide, identified, 111
Carbonization by heat, 96
Carmine flame, 212
Caseine, 159
Cast-iron identified, 53
Caustic, 26
  potash, 83

Caustic soda, 88
Cawk, 125
Celestine, 71, 125
Ceroleine, 159
Cerotic acid, 159
Cetine, 159
Chalcedony, 126
Chalk, 73
Char, 123
Charcoal, animal, 123
  blowpipe, 191
  wood, 123
Chili saltpetre, 88
Chloral, hydrate, 150
Chlorate of baryta, 70
  potash, 82
    blowpipe test, 220, 222
Chlorates, common, 98
  on charcoal, 194
  test for, 241
Chloric acid, 98
  detected, 91
Chloric peroxide, 97
Chloride, mercurous, 27
  of ammonium, NH$_4$Cl, 79
    for testing, 231
    use in analysis, 23
  of barium for testing, BaCl$_2$, 231
  of calcium for testing, CaCl$_2$, 231
  of lead, 30
  of lime, 74
    for testing, 231
  of mercury, HgCl$_2$, 27, 41
  of silver, 26
  of tin, 36
Chlorides detected, 91
  by blowpipe, 218
  impurities, 103
  test for, 237
Chlorine confirmed, 103
  detected, 91
  evolved, 29, 63
  preparation, 232
  water, for testing, 232
Chloroform, identified, 165
  tested, 165
Chloropicrine, 149
Cholesterine identified, 157
Chromate of lead, 29
  of potash, K$_2$CrO$_4$, 67
Chromates detected, 92

INDEX. 249

Chrome alum, 67
   iron ore, 66
   orange, 29
   yellow, 29
Chromic acid, 114
   detected, 92
   identified, 114
   test for, 228
Chromium, 66
   blowpipe test for, 209
   common compounds of, 66
   confirmed, 66
      by blowpipe, 209
   detected, 51
   insoluble compounds, 128
   oxide, 66
   test for, 239
Cinchonine detected, 144
      in quinine, 147
   identified, 148
   sulphate, 148
Cinnabar, 41
Citrates, common, 140
Citric acid detected, 130
   identified, 139
Classification of metals, 18
Clay, 59, 127
Cleavage, 31
Coal identified, 123
Coal-tar odor, 148
Cobalt, 57
   blowpipe test, 209
   confirmed, 57
   detected, 51
   glance, 57
   metallic, 57
   nitrate, $Co(NO_3)_2$, 57
   ores, 209
   oxide, 57
   test for, 239
   test in blowpipe analysis, 196
Cobaltous nitrate, 236
Coke identified, 123
Colcothar, 55
Colored beads, 208
   flames, 212
   flame test, 74
Common salt, 88
Concentrated sulphuric acid, H_2SO_4, 241
Condenser, 168
Condy's disinfectant, 63

Copper, 36
   acetate, 37
   alloys, 204
   arsenite, 37, 45
   blowpipe test, 210, 212
   carbonate, 37
   common compounds of, 37
   confirmed, 36
   detected, 32
   detected in lead, 206
   detection by blowpipe, 203
   ferrocyanide, 111
   glance, 38
   identified, 36
   metals precipitated by, 232
   ore, gray, 204
   oxide, 38
   oxychloride, 38
   pyrites, 203
   suboxide, 38
   sulphate, $CuSO_4$, 37
   sulphide, 38
   test for, 229, 233
   test for nitric acid, 105
Copperas, 55
Cork borers, 171
Corks fitted, 172
   perforated, 171
Corrosive sublimate, 31
Coughing indicates succinic acid, 132
Cream of tartar, 83
Crimson flame, 71
Crocus, 55
Crucible tongs, 117
Crushing-mortar, 120
Cryolite, 97
Crystals formed on cooling, 30
Cupellation, 205
Cupric oxide, 38
   sulphate, $CuSO_4$, 240
Cuprous oxide, 38
   sulphide, 38
Cyanide of mercury, 41
   analysis, 101
   potassium, commercial, KCN, 84
      for testing, 232
      in blowpipe analysis, 207
Cyanides, common, 101
   detected, 93
Cyanogen, 222
   detected, 93

DECANTATION, 33
  Decrepitation, 30, 41, 65
Deflagration, 82
Deliquescence, 62, 74, 82
Dextrine identified, 155
Diamond, 122
  mortar, 120
Dilute hydrochloric acid for testing, 233
  nitric acid for testing, 237
Dilute sulphuric acid for testing, 240
Dinas fire-brick, 127
Dish, evaporating, 94
Disinfectant, Burnett's, 62
  Condy's, 63
Dissolving, 19
Distillate, 169
Distillation, 168
Distilled water, 241
Dolomite, 73
Dropping-tube, 196

EARTHENWARE, 127
  Effervescence, 91
Efflorescence, 87
Electric calamine, 128
Emery, 59
Emulsion, 156
Epsom salts, 24
Essence of mirbane, 166
Ether, caution in using, 158
  for testing, 232
  identified, 165
Evaporating dish, 94
Evaporation, 94
  on glass, 20
  on water-bath, 174
Excess, meaning of, 21
Explosion with sulphuric acid, 97

FERRIC acetate, 134
  chloride, Fe_2Cl_6, 56
  citrate, 140
  oxide, 54
  salts detected, 52
Ferridcyanide of potassium, K_3FeCy_6, 84
  potassium, for testing, 232
Ferridcyanides detected, 100
Ferrocyanide of calcium and potassium, 111

Ferrocyanide of potassium, $K_4Fe Cy_6$, 84
Ferrocyanides detected, 110
  test for, 238
Ferrocyanogen, 110
Ferrous carbonate, 56
  iodide, 56
  oxide, 52
  salts detected, 52
Ferrous sulphate, $FeSO_4$, 55
  sulphide, 55
Fibrous gypsum, 73
File, rat's-tail, 171
Filters, 20
Filtration, 20
Fire-brick, 127
Fire-clay, 127
Flake white, 39
Flame, colored, test by, 74
Flask for distillation, 170
Flint, 126
Fluorescence, 147
Fluoric acid, 97
Fluoride of calcium, 65
Fluorides, blowpipe test, 219
  detected, 91
Fluorine, 97
  confirmed, 96
  detected, 91
Fluor-spar, 65
  blowpipe test, 214
Focus of blowpipe-flame, 203
Formic acid detected, 130
  identified, 134
French chalk, 128
Frosted silver, 206
Fructose, 153
Fruit-sugar, 153
Fuller's earth, 127
Fulminate of mercury, 101
Fumaric acid, 141
Fumes, cause of, 96
Fuming sulphuric acid, 102
Funnel-tube, 170
Fusing-point determined, 157
Fusion of insoluble substances, 120

GALENA, 30
  Gallic acid detected, 130
    identified, 135
    test for, 238
Garlic odor of arsenic, 45

Gas blowpipe, 173
Gas-burners, 94
Gas-carbon identified, 123
Gauze-burner, 95
Gelatine identified, 155
German tubing, 33
Glacial phosphoric acid, 107
Glance cobalt, 57
Glass jet, 43
    of borax, 90
    of metaphosphate of soda, 108
    rod, 22
    soluble, 90
    tubes bent, 172
    with carbonate of soda, 224
Glauber's salt, 88
Glucose identified, 152
Glycerin identified, 174
Gold, Mosaic, 47
    test for, 240
Goulard's extract, 133
Granulated zinc, 242
Grape-sugar identified, 152
Graphite, 123
Green acetate of copper, 37
    Brunswick, 38
    carbonate of copper, 37
    flame, 109, 212
    fused mass, 124
    nitric acid, 106
    oxide of chromium, 66
    Scheele's, 37, 45
    solution, 36, 58, 64, 67, 84
    vitriol, 55
Gray antimony ore, 49
    copper ore, 204
Grough saltpetre, 82
Gum identified, 155
    British, 155
Gun-metal, 204
Gunpowder, analysis of, 244
Gypsum, 73
    blowpipe test, 213

HÆMATITE, 54
    Hartshorn, 79
Heating solids in air, 117
    in bent tube, 118
    in tubes, 33
    on platinum foil, 117

Heavy spar, 69, 125
    blowpipe test, 213
Hippuric acid detected, 130
    identified, 136
Hydriodate of potash, 85
Hydriodic acid detected, 92
    test for, 237
Hydrobromic acid detected, 96
Hydrochloric acid, HCl, 104
    detected, 91
    for testing, 233
    identified, 104
    precipitate, 25
    test for, 237
Hydrocyanic acid confirmed, 100
    detected, 93
    identified, 101
Hydrodisodic phosphate, 238
Hydroferrocyanic acid identified, 110
Hydrofluoric acid detected, 91
    identified, 97
Hydrofluosilicic acid for testing, $2HF.SiF_4$, 233
Hydrogen prepared, 43
    tested, 43
Hydropotassic metantimoniate, 229
    · sulphate, 230
Hydrosodic carbonate, 87
    for testing, 229
Hydrosulphate of ammonia, NH_4HS, 80, 233
    precipitate, 51
    yellow, 80
Hydrosulphocyanic acid, 101
Hydrosulphuric acid, H_2S, 233
    apparatus, 234
    confirmed, 99
    detected, 91
    for testing, 233
    identified, 99
    in excess, 21
    precipitate, 32
    prepared, 234
    test for, 228
Hypochlorite of lime, 74
Hypochlorites, test for, 240
Hypochlorous acid confirmed, 98
    detected, 93
Hyposulphite of soda, 89
Hyposulphites detected, 91

ICELAND SPAR, 73
  Impalpable powder, 120
Incandescence, 208
Incense, odor of, 132
Incrustation on charcoal, 192
Indigo, action of heat on, 149
  identified, 149
  reduced, 149
Ink identified, 135
Insoluble substances analyzed, 116
Iodide of arsenic, 46
  of lead, 30
  of mercury, 41
  of potassium for testing, KI, 235
Iodides, blowpipe test, 217
  common, 99
  detected, 91
  test for, 237
Iodine confirmed, 98
  detected, 91
  for testing, 235
  free test for, 240
  identified, 98
    by blowpipe, 221
  vapor, 221
  water, 235
Iron acetate, 134
  ammonio-citrate, 140
  and quinine citrate, 140
  as an impurity, 210
  bisulphide, 55
  black oxide, 55
  blowpipe test, 208
  carbonate, 56
  cast, identified, 53
  citrate, 140
  common compounds of, 54
  confirmed, 52
  detected, 51
    by blowpipe, 208
  ferrocyanide, 56, 111
  identified, 52
  iodide, 56
  magnetic oxide, 55
  muriate, 56
  perchloride, Fe_2Cl_6, 56
  peroxide, 54
  phosphate, 109
  protosulphate, $FeSO_4$, 55
  pyrites, 55
  sesquichloride, Fe_2Cl_6, 56

Iron, sesquioxide, 54
  silicate, 56
  slag, 56
  spathic ore of, 56
  specular, 54
  sulphate, $FeSO_4$, 55
    action of heat on, 210
  sulphide, 55
  sulphuret, 55
  tannate, 135
  test for, 52, 229, 233
  tinned, 207
  wire for blowpipe analysis, 226
  wrought, identified, 53
Ivory black, 123

JET for burning gases, 43
  Joints for apparatus, 168

KAOLIN, 127
  Kaolinite, 127
Kryolite, 97
  identified, 97

LACTIC acid identified, 174
  Lactide, 174
Lactine identified, 153
Lamp-black identified, 123
Lamps, 94
Lead, 28, 204
  acetate, $Pb(C_2H_3O_2)_2$, 29
    for testing, 228
  alloys, 205
  binoxide, 29
  blowpipe test, 201
  carbonate, 29
  chloride, 30
  chromate, 29
  common compounds of, 28
  confirmed, 27
  detected, 25, 32
  extracted from galena, 204
  identified, 28
  iodide, 30
  nitrate, 30
  ore, 204
  oxide, 28
  oxychloride, 30
  peroxide, 29
  phosphate, 110
  precipitated by sulphuric acid, 27

INDEX. 253

Lead, red oxide, 29
  sugar of, 29
  sulphate, 30
  sulphide, 30
  sulphide in blowpipe analysis, 204
  tartrate, 138
  test for, 235
  tested for copper, 206
  tested for silver, 205
  tribasic acetate, 133
  white, 29
Lemons, essential salt of, 85
Levigation in blowpipe analysis, 193
Liebig's condenser, 168
Lime, CaO, 72
  blowpipe test, 212
  borate, 113
  carbonate, 72
  chloride, 74
  citrate, 141
  detected, 68
  hydrate, $CaO.H_2O$, 72
  hypochlorite, 74
  oxalate, 65, 74
  phosphate, 64
    blowpipe test, 226
    detected, 51
  slaked, 72
  sulphate, $CaSO_4$, 73
    blowpipe test, 213
  superphosphate, 64
  tartrate, 137
  water, $CaO.H_2O$, 72
Limestone, 73
Liquids, unknown, examined, 181
Liquor ammoniæ, 79
Liquor potassæ, 83
Litharge, 28
Lithates, 142
Lithic acid identified, 141
Lunar caustic, 26

MAGNESIA, 24
  ammonia phosphate, 65
  basic carbonate, 24
  calcined, 24
  carbonate, 24
  citrate, 141
    granulated, 141
  common compounds of, 23

Magnesia detected, 17
  phosphate, 65
  precipitated by ammonia, 23
  silicate, 128
  sulphate, $MgSO_4$, 24
Magnesian limestone, 73
Magnesic sulphate, 24, 240
Magnesite, 24
Magnesium blowpipe test, 190
  common compounds of, 23
  detected, 17
  identified, 23
Malachite, 37
Malæic acid, 141
Mallic acid detected, 130
  identified, 141
Malleability tested, 192
Manganate of potash, 64
  Manganese, 62
  binoxide, MnO_2, 63
  black, 63
  blowpipe test, 208
  common compounds of, 63
  confirmed, 62
    by blowpipe, 209
  detected, 51
  ore, 63
  oxide, 63
  sulphate, $MnSO_4$, 63
  test for, 239
Manganous sulphate, 63, 240
Marble, 73
Marsh's test for arsenic, 42
Massicot, 28
M.B. examination, organic substances for, 183
Meconic acid detected, 130
  identified, 135
Meerschaum, 128
Melting-point determined, 157
Mercuric chloride, $HgCl_2$, 41, 238
  compounds, 40
  cyanide, 41
  iodide, 41
  oxide, 41
  sulphide, 41
Mercurous chloride, 27
  compounds, 27
  nitrate, 27
Mercury, 27
  acetate, 133
  bichloride, $HgCl_2$, 41
  blowpipe test, 190

Mercury, chloride, $HgCl_2$, 41
   common compounds of, 27, 40
   confirmed, 40
   cyanide, 41
   detected, 25, 32
   fulminate, 101
   iodide, 41
   nitric oxide, 41
   oxide, 41
   perchloride, $HgCl_2$, 41
   protochloride, 27
   protonitrate, 27
   red oxide, 41
   subchloride, 27
   sulphide, 41
   test for, 240
Metals detected by blowpipe, 190
   by borax beads, 208
   by colored flames, 212
Metaphosphoric acid, 107
Methylated finish, 228
   spirit, 162
Methylic alcohol identified, 62
Microcosmic salt, 108
   bead, 218
   for testing, 236
Milkiness with water, 29, 38, 49
Milk-sugar identified, 153
Millon's test, 160
Mine tin ore, 125
Minium, 29
Molybdate of ammonia for testing, 236
Molybdate of ammonia test, 107
Morphine, acetate, 146
   detected, 144
   hydrochlorate, 146
   identified, 145
   meconate, 145
   muriate, 146
Mortar, 118
Mosaic gold, 47
Murexide test for uric acid, 131

NAPHTHALINE identified, 158
   Narcotine detected, 144
   identified, 147
Needle-like crystals, 28
Nickel, 57
   blowpipe test, 208
   confirmed, 57
   by blowpipe, 211

Nickel detected, 51
   ores, 212
   oxide, 58
   speiss, 212
   sulphate, 58
   test for, 229, 239
Nicotine, identified, 163
Nitrate, mercurous, 27
Nitrate of baryta, $Ba(NO_3)_2$, 236
   of bismuth, 39
   of cobalt, $Co(NO_3)_2$, 236
   of lead, 30
   of potash for testing, KNO_3, 236
   of silver, $AgNO_3$, 26
   for testing, 236
Nitrates, action of heat on, 222
   blowpipe test, 217
   on charcoal, 223
Nitre, KNO_3, 81
   blowpipe test, 219
   cubic, 88
Nitric acid, HNO_3, 105
   confirmed, 105
   detected, 92
   for testing, 237
   identified, 105
Nitrites detected, 105
Nitrobenzole, 166
Nitroprusside of sodium, 237
   for testing, 237
Nitrous acid, commercial, 106
   detected, 105
Non-metals detected, 91

OCCLUSION of oxygen by silver, 206
Octahedra of arsenious acid, 42
Odors of organic acids, 132
Oil of bitter almonds, 166
   vitriol, 102
   identified, 102
Oleic acid, 174
Oleine, 174
Orange chrome, 29
Organic acid detected in alkaline solution, 131
   detected in aqueous solution, 130
   detected in insoluble substance, 131
   acids detected, 130
   by odor, 132

Organic acids, salts of, analyzed, 132
  bases identified, 144
  liquids miscible with potash, 164
    with water, 162
    not miscible with hydrochloric acid, potash, or water, 165
  matter detected, 96
  solids tested, 149
  substances defined, 96
  substances, acid, 179
    bitter, 179
    dissolved by alcohol, 156
    dissolved by boiling water, 154
    dissolved by cold water, 152
    dissolved by ether, 158
    evolving ammonia, 177
    for M.B. examination, 183
    fusible, 177
    identified, 149
    insoluble, 159
      in boiling water, 179
      in cold water, 178
    liquid, 161
    nitrogenized, 177
    soluble in potash, 185
    sweet, 186
    unknown examination, 176
    volatile, 177
Original solution, 17
Orpiment, 46
Orthophosphoric acid, 107
Oxalate of ammonia $(NH_4)_2C_2O_4$, 237
  of lime, 65, 74
Oxalic acid confirmed, 111
  detected, 92
  identified, 111
Oxidizing flame, 191
Oxygen evolved, 82

PALMITIC acid identified, 157
  Palmitine identified, 158
Paraffine identified, 159
Peacock ore, 203

Pearlash, 82
Pearl white, 39
Perchloride of iron, Fe_2Cl_6, 237
  mercury, $HgCl_2$, 238
Perforated corks, 171
Permanganate of potash, $KMnO_4$, 63
Peruvian saltpetre, 88
Pestle and mortar, 118
Pewter, 205
Phenic acid identified, 150
Phenole identified, 150
Phosphate of lime detected, 64
  of magnesia, 65
  of magnesia and ammonia, 65, 109
  of soda, Na_2HPO_4, 89
    and ammonia, 108
    for testing, 238
Phosphates, 103
  blowpipe test, 226
Phosphorescence, 65
Phosphoric acid, blowpipe test, 226
  confirmed, 106
  detected, 92
  identified, 107
  test for, 107
  tribasic, 107
Phosphorus, amorphous, identified, 108
  salt, 108
  vitreous, identified, 107
Picric acid identified, 149
Pig-iron identified, 53
Pineapple odor, 164
Pink salt, 48
  solution, 57
  sulphate of manganese, 63
Pipe-clay, 127
Plaster of Paris, 73
Platina, muriate, 229
Platinic chloride, $PtCl_4$, 229
Platinum, bichloride, $PtCl_4$, 229
  capsule, 121
  chloride, $PtCl_4$, 229
  corroded, 125
  foil, 120
    for fusing, 120
  test for, 231, 240
  wire cleaned, 195
    for borax beads, 194
    colored flames, 78

Plumbago, 123
Plumbic acetate, 228
Porcelain, 127
Potash, KHO, 83
  acetate, 134
  alum, 59
  antimoniate, $KSbO_3$, 50
  bicarbonate, 83
  bichromate, $K_2O,2CrO_3$, 67
  binoxalate, 85
  bisulphate, $KHSO_4$, 83
    blowpipe test, 217
  bitartrate, 83, 138
    blowpipe test, 222
  carbonate, 82
  caustic, 83
  chlorate, 82
  chromate, 67
  common compounds of, 81
  for testing, KHO, 238
  hydrate, 83
  hydriodate, 85
  manganate, 64
  nitrate, KNO_3, 81
  oleate, 85
  oxalate, 85
  permanganate, $KMnO_4$, 63
  prussiate, 84
  red prussiate, 84
  silicate, 85
  solution of, KHO, 83
  sulphate, 83
Potashes, American, 82
Potassa fusa, 83
Potassic cyanide, KCN, 84, 232
  dichromate, 67, 229
  ferridcyanide, 84, 232
  ferrocyanide, 84, 232
  hydrate, KHO, 83, 238
  iodide, KI, 85, 235
  nitrate, KNO_3, 81, 236
Potassium, 81
  blowpipe test, 212
  carbonate, 82
  chloride, 83
  common compounds of, 81
  cyanide, KCN, 84, 232
  detected, 77
  ferridcyanide, K_3FeCy_6, 84, 232
  ferrocyanide, K_4FeCy_6, 84, 232
    blowpipe test, 214

Potassium, iodide, KI, 85
  nitrate, KNO_3, 81
  sulphate, 83
  sulphide, 99
  sulphocyanide, 101
Powdering substances, 118
Precipitate defined, 17
  washed, 33
  white, 41
Precipitation promoted by stirring, 22, 78
Prepared chalk, 73
Preston salts, 80
Protochloride of tin for testing, $SnCl_2$, 239
Protosulphate of iron, $FeSO_4$, 240
Prussian blue, 52, 56, 111
  test, 100
Prussiate of potash, 84
Prussic acid, 101
  detected, 93
  identified, 101
Pulverization, 118
Pumice-stone, 128
Purple crystals, 67
  solution, 63
  vapors, 27, 94
Putty powders, 125
Pyrites, 55
  copper, 203
  iron, 55
Pyrogallic acid identified, 153
Pyrogalline, 153
Pyroligneous acid, 133
  ether, 162
Pyrolusite, 63
Pyroxylic spirit, 162

QUARTZ, 126
Quicklime, 72
Quicksilver, 27
Quinine and iron citrate, 140
  detected, 144
  identified, 147
  sulphate, 147

RAIN-WATER, 242
Rat's-tail file, 171
Reagents, 228
Realgar, 46
Red chlorosulphide of lead, 32
  chromate of lead, 29
  chromate of potash, 67

INDEX. 257

Red color with sulphuric acid, 154
  drops, 221
  flame, 71, 212
  hæmatite, 54
  iodide of arsenic, 46
  iodide of mercury, 41
  lead, 29
  litmus paper, 235
  mordant, 60
  nitroprusside of sodium, 237
  orpiment, 46
  oxide of copper, 38
    iron, 54
    lead, 29
    mercury, 41
  phosphorus, 108
  precipitate, 41
  prussiate of potash, 84
  solution, 63, 67, 134
  sulphide of antimony, 49
    mercury, 41
  vapors, 91, 103
Reducing flame, 191
Reduction of metals on charcoal, 191
Reinsch's test for arsenic, 42
Retort, 168
Retort stand, 94
Ring gas-burner, 169
Roasting before blowpipe, 203
Rochelle salt, 139
Rod, breaking and rounding, 22
Roll brimstone, 122
Rose gas-burner, 169
Rosin identified, 157
Rouge, jeweller's, 55
Rust, 55

SAL-AMMONIAC, 79
  Salicine detected in quinine, 147
    identified, 154
Saliretine, 154
Sal prunella, 82
Salt, 88
  cake, 88
  common, 88
  of lemon, 85
  of sorrel, 85
  of tartar, 82
Saltpetre, 81
  flour, 82

Saltpetre, Peruvian, 88
Sand, 126
Scheele's green, 37, 45
Selenite, 73
Shrapnel bullets, 205
Sifting, 119
Silica, 126
  amorphous, 126
  detected, 113
    by blowpipe, 224
  soluble, 126
Silicic acid, 126
  detected, 113
    by blowpipe, 224
  separation from solutions, 113
Silicates, common simple, 127
  detected, 113
  on charcoal, 223
  soluble, 114
Silicium, 126
Silicofluoric acid, 233
Silicon, 126
  fluoride, 97
  oxide, 126
Silver acetate, 133
  blowpipe test, 201, 206
  chloride, 26, 103, 124
    in blowpipe analysis, 206
  common compounds of, 26
  cyanide, 103
  detected, 25
  extracted from lead, 205
Silver ferridcyanide, 110
  ferrocyanide, 110
  identified, 26
  nitrate, $AgNO_3$, 26
  oxalate, 112
  precipitates distinguished, 103
  test for, 235
  tested for copper, 206
Singed hair, odor of, 132
Slag, iron, 56
Slaked lime, 72
Slate, 127
Smalt, 57
Smelling salts, 80
Smell of almonds, 132, 166
  burnt sugar, 137
  chlorine, 74
  gas, 158, 166

Smell of incense, 136
  singed hair, 132
  tar, 164
Soap, 90
  soda, 90
  soft, 85
Soap-stone, 128
Soda, acetate, 134
  aluminate, 60
  arseniate, 46
  ash, 88
  biborate, $Na_2O.2B_2O_3$, 89
    blowpipe test, 226
  bicarbonate, $NaHCO_3$, 87
  carbonate, Na_2CO_3, 87
    dried, 230
    for blowpipe analysis, 193
  caustic, 88
  chloride, 89
  crystals, 87
  hydrate, 88
  hyposulphite, 89
    blowpipe test, 215
  nitrate, 88
    blowpipe test, 215
  orthophosphate, 89
  phosphate, Na_2HPO_4, 89
  silicate, 90
  stannate, 47
  sulphate, 88
  sulphite, 89
  tungstate, 90
  urate, 142
  waste, 100
Sodic carbonate, 87, 230
Sodium, blowpipe test, 212
  carbonate, 87
  chloride, 88
  common compounds of, 87
  detected, 77
  flame deceptive, 214
  hyposulphite, 89
  impurity, 214
  nitrate, 88
  phosphate, 89
  sulphate, 88
  sulphite, 89
Soft soap, 85
Solder, 205
Solid unknown substances examined, 181
Soluble glass, 90

Solution, 19
Soot identified, 124
Spathic iron ore, 56
Spatula, 118
Specular iron ore, 54
Speiss, 212
Spelter (zinc), 60
Spermaceti identified, 159
Spirit-black, 123
Spirit-blowpipe, 173
Spirit-lamp, 19
Spirit of wine, 162
Spirting, prevention of, 94
Spongy flakes of sulphur, 41
Stannate of soda, 47
Stannic acid, 47, 125
  chloride, 48
  compounds, 47
Stannous chloride, $SnCl_2$, 36, 234
  compounds, 36
Starch for testing, 240
  identified, 154
  sugar, 152
  test for, 235
Steam-bath for evaporation, 174
Stearic acid identified, 157
Stearine identified, 156
Steatite, 128
Steel identified, 53
Stirring, 22
  rod, 22
  to promote precipitation, 22, 78
Stoppers, India-rubber, 172
Stourbridge clay, 127
Stream tin ore, 125
Strong sulphuric acid, H_2SO_4, 241
Strontia, carbonate, 71
  detected, 68
  nitrate, 71
  sulphate, 125
Strontianite, 71
Strontium, 70
  blowpipe test, 212
  confirmed, 70
  detected, 68
Strychnine detected, 144
  identified, 146
Sublimate corrosive, 41
Sublimation, 136
Succinic acid detected, 130
  identified, 136

INDEX.

Sugar, 152
    copper test for, 152
    identified, 152
    of lead, 29
Sulphate of copper, $CuSO_4$, 37
    for testing, 240
    of iron for testing, $FeSO_4$, 240
    lead, 30
    of lime for testing, $CaSO_4$, 240
    of magnesia, $MgSO_4$, 24
    for testing, 240
    of manganese for testing, $MnSO_4$, 240
    of zinc, 61
Sulphates, action of heat on, 221
    blowpipe test, 224
    detected, 92
    impurities, 102
    insoluble, 125
Sulphide of ammonium, $(NH_4)_2S$, 80
    for testing, 233
    of ammonium precipitate, 51
    of antimony, 49
    of arsenic, 46
    of copper, 38
    of iron, 55
    of lead, 30
    of tin, 36
    of zinc, 61
Sulphides, action of heat on, 222
    common, 99
    detected, 91
        by blowpipe, 225
    on charcoal, 223
Sulphites detected, 91
Sulphocyanides detected, 92
    test for, 241
Sulphur, 122
    blowpipe test, 222
    crude, 122
    detected, 91
        by blowpipe, 222
    distilled, 122
    evolved, 222
    expelled by roasting, 203
    flowers of, 122
    identified, 122
    insoluble, 122
    milk of, 122
    precipitated, 122
    precipitation of, 23

Sulphur roll, 122
    Sicilian, 122
    soluble, 122
    sublimed, 122
    viscous, 122
Sulphurets, common, 99
Sulphuretted hydrogen apparatus, 234
    identified, 98
    precipitate, 32
    prepared, 233
    test for, 228
Sulphuric acid, H_2SO_4, 102
    blowpipe test, 224
    detected, 92
    for testing, 241
    identified, 102
    Nordhausen, 102
    test for, 236
Sulphurous acid confirmed, 101
    detected, 91
    evolved, 223
Sulphydrate of ammonium, 233
Superphosphate of lime, 64
Sweet substances, 186
Sweet taste, 29
Sylvic acid, 158

TABLE A, page 17
    " B, ... 25
    " C, ... 32
    " D, ... 51
    " E, ... 68
    " F, ... 77
    " G, ... 91
    " H, ... 92, 93
    " I, ... 116
    " K, ... 129
    " L, ... 130
    " M, ... 131
    " N, ... 131
    " O, ... 144
    " P, ... 151
    " Q, ... 161
    " R, ... 190
    " S, ... 201
    " T, ... 208
    " U, ... 212
    " V, ... 216
    " W, ... 217
    " X, ... 221
    " Y, ... 223
    " Z, ... 224

Tannic acid detected, 130
    identified, 135
    test for, 238
Tannin, 135
Tartar, cream of, 83
  emetic, 49
    blowpipe test, 202
  salt of, 82
Tartaric acid, $H_2C_4H_4O_6$, 137
  detected, 130
  for testing, 241
  identified, 137
Tartrates, action of heat on, 222
  common, 138
Test papers, 235, 241
  use of, 34
  tube, 19
    rack, 19
Tests, 228
  addition of, 21
Theine detected, 145
  identified, 145
Thermometer, 168
Tin, 35
  alloys of, 205
  before blowpipe, 207
  bichloride, 48
  bichloride with hydrochlorate of ammonia, 48
  binoxide, 47, 125
  bisulphide, 47
  blowpipe test, 201
  common compounds of, 36, 47
  confirmed, 35, 46
  crystals, 36
  detected, 32
    by blowpipe, 201
  foil, 239
  granulated, 239
  identified, 35
  in insoluble substances, 125
  nitromuriate, 48
  ore, 125
  persalts of, 47
  plate, 207
  protochloride, $SnCl_2$, 36
  protosalts of, 36
  protosulphide, 36
  pyrites, 36
  reduction on charcoal, 190
  salts of, 36, 47
  stone, 125

Tin test for, 238
Tincture of iron, 56
Toluidine, 164
Tongs, 117
Triangle, 117
Triple phosphate, 65, 109
Tube, blowpipe test in, 221
  funnel, 170
  German, 33
Tubes, bent, 172
  sealed, 33
Tubulated retort, 168
Tubulus, 168
Tungstate of soda, blowpipe test, 215
Tungstic acid detected, 90
Turmeric, 241
  paper, 241
Type-metal, 205

ULTRAMARINE, 100
  University of London examination, 183
Unknown liquid examined, 181
  solid examined, 181
Urates, common, 142
Urea identified, 153
  nitrate, 155
  oxalate, 155
Uric acid confirmed, 141
  detected, 131
  identified, 141

VALERIANIC acid, 164
Verdigris, 37
Vermilion, 41
  antimony, 49
Vinegar, 133
Violet flame, 81
Vitriol, blue, 37
  green, 55
  identified, 102
  white, 61

WASHING bottle, 33
  precipitates, 33
Washing-soda, 87
Water, H_2O, 181
Water-bath, 174
Water for testing, 241
  glass, 90
  purification, 242
  tested for impurities, 242

Wax, identified, 159
White lead, 29
    precipitate, 41
    vitriol, 61
Wire triangle, 117
Witherite, 69
Wood-charcoal, 123
Wood-naphtha, 162
Wood-spirit, 163

YELLOW chromate of baryta, 70
    of potash, 67
  chrome, 29
  flame, 212
  fused mass, 124
  iodide of lead, 30
    of mercury, 41
  orpiment, 46
  oxide of lead, 28
  oxychloride of lead, 30
  picric acid, 149

Yellow prussiate of potash, 84
    sulphide of tin, 47

ZEOLITES, 114
  Zeolitic minerals, 114
Zinc before blowpipe, 208
  blowpipe test, 190
  carbonate, 61
  chloride, 62
  common compounds of, 61
  confirmed, 60
  detected, 51
  glance, 128
  granulated, 43
  identified, 60
  lactate, 174
  oxide, 61
  silicate, 128
  sulphate, 61
  sulphide, 61
  test for, 235
  white, 61

CATALOGUE No. 7.   MAY, 1889.

A CATALOGUE

OF

BOOKS FOR STUDENTS.

INCLUDING THE

?QUIZ-COMPENDS?

CONTENTS.

	PAGE		PAGE
New Series of Manuals,	2,3,4,5	Obstetrics,	10
Anatomy,	6	Pathology, Histology,	11
Biology,	11	Pharmacy,	13
Chemistry,	6	Physical Diagnosis,	11
Children's Diseases,	7	Physiology,	12
Dentistry,	8	Practice of Medicine,	12
Dictionaries,	8	Prescription Books,	12
Eye Diseases,	8	?Quiz-Compends?	15, 16
Electricity,	9	Skin Diseases,	13
Gynæcology,	10	Surgery,	13
Hygiene,	9	Therapeutics,	9
Materia Medica,	9	Throat,	14
Medical Jurisprudence,	9	Urine and Urinary Organs,	14
Miscellaneous,	10	Venereal Diseases,	14

PUBLISHED BY

P. BLAKISTON, SON & CO.,

Medical Booksellers, Importers and Publishers.

LARGE STOCK OF ALL STUDENTS' BOOKS, AT
THE LOWEST PRICES.

1012 Walnut Street, Philadelphia.

*** For sale by all Booksellers, or any book will be sent by mail, postpaid, upon receipt of price. Catalogues of books on all branches of Medicine, Dentistry, Pharmacy, etc., supplied upon application.

"*An excellent Series of Manuals.*"—*Archives of Gynæcology*

A NEW SERIES OF
STUDENTS' MANUALS

On the various Branches of Medicine and Surgery.

Can be used by Students of any College.

Price of each, Handsome Cloth, $3.00. Full Leather, $3.50.

The object of this series is to furnish good manuals for the medical student, that will strike the medium between the compend on one hand and the prolix textbook on the other—to contain all that is necessary for the student, without embarrassing him with a flood of theory and involved statements. They have been prepared by well-known men, who have had large experience as teachers and writers, and who are, therefore, well informed as to the needs of the student.

Their mechanical execution is of the best—good type and paper, handsomely illustrated whenever illustrations are of use, and strongly bound in uniform style.

Each book is sold separately at a remarkably low price, and the immediate success of several of the volumes shows that the series has met with popular favor.

No. 1. SURGERY. 236 Illustrations.

A Manual of the Practice of Surgery. By WM. J. WALSHAM, M.D., Asst. Surg. to, and Demonstrator of Surg. in, St. Bartholomew's Hospital, London, etc. 228 Illustrations.

Presents the introductory facts in Surgery in clear, precise language, and contains all the latest advances in Pathology, Antiseptics, etc.

"It aims to occupy a position midway between the pretentious manual and the cumbersome System of Surgery, and its general character may be summed up in one word—practical."—*The Medical Bulletin.*

"Walsham, besides being an excellent surgeon, is a teacher in its best sense, and having had very great experience in the preparation of candidates for examination, and their subsequent professional career, may be relied upon to have carried out his work successfully. Without following out in detail his arrangement, which is excellent, we can at once say that his book is an embodiment of modern ideas neatly strung together, with an amount of careful organization well suited to the candidate, and, indeed, to the practitioner."—*British Medical Journal.*

Price of each Book, Cloth, $3.00; Leather, $3.50.

No. 2. DISEASES OF WOMEN. 130 Illus.
The Diseases of Women. By DR. F. WINCKEL, Professor of Gynæcology and Director of the Royal University Clinic for Women, in Munich. Translated from the German by DR. J. H. WILLIAMSON, Resident Physician Allegheny General Hospital, Allegheny, Penn'a, under the supervision of, and with an introduction by, **Theophilus Parvin**, M.D., Professor of Obstetrics and Diseases of Women and Children in Jefferson Medical College. Illustrated by 132 fine Engravings on Wood, most of which are new.

"The book will be a valuable one to physicians, and a safe and satisfactory one to put into the hands of students. It is issued in a neat and attractive form, and at a very reasonable price."—*Boston Medical and Surgl. Journal.*

No. 3. OBSTETRICS. 227 Illustrations.
A Manual of Midwifery. By ALFRED LEWIS GALABIN, M.A., M.D., Obstetric Physician and Lecturer on Midwifery and the Diseases of Women at Guy's Hospital, London; Examiner in Midwifery to the Conjoint Examining Board of England, etc. With 227 Illus.

"This manual is one we can strongly recommend to all who desire to study the science as well as the practice of midwifery. Students at the present time not only are expected to know the principles of diagnosis, and the treatment of the various emergencies and complications that occur in the practice of midwifery, but find that the tendency is for examiners to ask more questions relating to the science of the subject than was the custom a few years ago. * * * The general standard of the manual is high; and wherever the science and practice of midwifery are well taught it will be regarded as one of the most important text-books on the subject."—*London Practitioner.*

No. 4. PHYSIOLOGY. Third Edition.
321 ILLUSTRATIONS AND A GLOSSARY.
A Manual of Physiology. By GERALD F. YEO, M.D., F.R.C.S., Professor of Physiology in King's College, London. 321 Illustrations and a Glossary of Terms. Third American from second English Edition, revised and improved. 758 pages.

This volume was specially prepared to furnish students with a new text-book of Physiology, elementary so far as to avoid theories which have not borne the test of time and such details of methods as are unnecessary for students in our medical colleges.

"The brief examination I have given it was so favorable that I placed it in the list of text-books recommended in the circular of the University Medical College."—*Prof. Lewis A. Stimson,* M.D., *37 East 33d Street, New York.*

Price of each Book, Cloth, $3.00; Leather, $3.50.

No. 5. POTTER'S MATERIA MEDICA, PHARMACY AND THERAPEUTICS.

OVER 600 PRESCRIPTIONS, FORMULÆ, ETC.

A Handbook of Materia Medica, Pharmacy and Therapeutics—including the Physiological Action of Drugs, Special Therapeutics of Diseases, Official and Extemporaneous Pharmacy, etc., etc. By SAM'L O. L. POTTER, M.A., M.D., Professor of the Practice of Medicine in Cooper Medical College, San Francisco, Late A. A. Surg., U. S. A., Author of the "Quiz-Compends" of Anatomy and Materia Medica, etc.

This book contains many unique features of style and arrangement; no time or trouble has been spared to make it most complete and yet concise in all its parts. It contains 600 prescriptions of practical worth, a great mass of facts conveniently and concisely put together, also many tables, dose lists, diagnostic hints, etc., all rendering it the most complete manual ever published.

☞ Part III, Special Therapeutics, consists of an Alphabetical List of Diseases, in which is given the proper drugs to be used in the treatment of each, with the authority recommending them, and in many cases signed prescriptions. This will be found of great value to the young practitioner, and to the physician of experience it will suggest new methods of treatment in obstinate and chronic cases.

"Dr. Potter's handbook will find a place, and a very important one, in our colleges and the libraries of our practitioners."—*N. Y. Medical Journal.*

No. 6. DISEASES OF CHILDREN.

A Manual. By J. F. GOODHART, M.D., Phys. to the Evelina Hospital for Children; Asst. Phys. to Guy's Hospital, London. American Edition. Edited by LOUIS STARR, M.D., Clinical Prof. of Dis. of Children in the Hospital of the Univ. of Pennsylvania, and Physician to the Children's Hospital, Phila. Containing many new Prescriptions, a list of over 50 Formulæ, conforming to the U. S. Pharmacopœia, and Directions for making Artificial Human Milk, for the Artificial Digestion of Milk, etc.

"As it is said of some men, so it might be said of some books, that they are 'born to greatness.' This new volume has, we believe, a mission, particularly in the hands of the younger members of the profession. In these days of prolixity in medical literature, it is refreshing to meet with an author who knows both what to say and when he has said it. The work of Dr. Goodhart

Price of each Book, Cloth, $3.00 ; Leather, $3.50.

(admirably conformed, by Dr. Starr, to meet American requirements) is the nearest approach to clinical teaching without the actual presence of clinical material that we have yet seen."—*New York Medical Record.*

No. 7. PRACTICAL THERAPEUTICS.
FOURTH EDITION, WITH AN INDEX OF DISEASES.

Practical Therapeutics, considered with reference to Articles of the Materia Medica. Containing, also, an Index of Diseases, with a list of the Medicines applicable as Remedies. By EDWARD JOHN WARING, M.D., F.R.C.P. Fourth Edition. Rewritten and Revised. By DUDLEY W. BUXTON, M.D., Asst. to the Prof. of Medicine at University College Hospital.

"We wish a copy could be put in the hands of every Student or Practitioner in the country. In our estimation, it is the best book of the kind ever written."—*N. Y. Medical Journal.*

No. 8. MEDICAL JURISPRUDENCE AND TOXICOLOGY. New Ed.

By John J. Reese, M.D., Professor of Medical Jurisprudence and Toxicology in the University of Pennsylvania; President of the Medical Jurisprudence Society of Phila.; 2d Edition, Revised and Enlarged.

"This admirable text-book."—*Amer. Jour. of Med. Sciences.*

"We lay this volume aside, after a careful perusal of its pages, with the profound impression that it should be in the hands of every doctor and lawyer. It fully meets the wants of all students. . . . . He has succeeded in admirably condensing into a handy volume all the essential points."—*Cincinnati Lancet and Clinic.*

No. 9. ORGANIC CHEMISTRY.

Or the Chemistry of the Carbon Compounds. By Prof. VICTOR VON RICHTER, University of Breslau. Authorized translation, from the Fourth German Edition. By EDGAR F. SMITH, M.A., PH.D.; Prof. of Chemistry in University of Pennsylvania; Member of the Chem. Socs. of Berlin and Paris.

"I must say that this standard treatise is here presented in a remarkably compendious shape."—*J. W. Holland,* M.D., *Professor of Chemistry, Jefferson Medical College, Philadelphia.*

"This work brings the whole matter, in simple, plain language, to the student in a clear, comprehensive manner. The whole method of the work is one that is more readily grasped than that of older and more famed text-books, and we look forward to the time when, to a great extent, this work will supersede others, on the score of its better adaptation to the wants of both teacher and student."—*Pharmaceutical Record.*

Price of each Book, Cloth, $3.00; Leather, $3.50.

ANATOMY.

Holden's Anatomy. A manual of Dissection of the Human Body. Fifth Edition. Enlarged, with Marginal References and over 200 Illustrations. Octavo. Cloth, 5.00; Leather, 6.00
    Bound in Oilcloth, for the Dissecting Room, $4.50.

"No student of Anatomy can take up this book without being pleased and instructed. Its Diagrams are original, striking and suggestive, giving more at a glance than pages of text description. * * * The text matches the illustrations in directness of practical application and clearness of detail."—*New York Medical Record.*

Holden's Human Osteology. Comprising a Description of the Bones, with Colored Delineations of the Attachments of the Muscles. The General and Microscopical Structure of Bone and its Development. With Lithographic Plates and Numerous Illustrations. Seventh Edition. 8vo. Cloth, 6.00

Holden's Landmarks, Medical and Surgical. 4th ed.
    Cloth, 1.25

Heath's Practical Anatomy. Sixth London Edition. 24 Colored Plates, and nearly 300 other Illustrations. Cloth, 5.00

Potter's Compend of Anatomy. Fourth Edition. 117 Illustrations. Cloth, 1.00; Interleaved for Notes, 1.25

CHEMISTRY.

Bartley's Medical Chemistry. A text-book prepared specially for Medical, Pharmaceutical and Dental Students. With 40 Illustrations, Plate of Absorption Spectra and Glossary of Chemical Terms. Cloth, 2.50

** This book has been written especially for students and physicians. It is practical and concise, dealing only with those parts of chemistry pertaining to medicine; no time being wasted in long descriptions of substances and theories of interest only to the advanced chemical student.

Bloxam's Chemistry, Inorganic and Organic, with Experiments. Sixth Edition. Enlarged and Rewritten. Nearly 300 Illustrations. Cloth, 4.50; Leather, 5.50

Richter's Inorganic Chemistry. A text-book for Students. Third American, from Fifth German Edition. Translated by Prof. Edgar F. Smith, PH.D. 89 Wood Engravings and Colored Plate of Spectra. Cloth, 2.00

Richter's Organic Chemistry, or Chemistry of the Carbon Compounds. Translated by Prof. Edgar F. Smith, PH.D. Illustrated. Cloth, 3.00; Leather, 3.50

☞ *See pages 2 to 5 for list of Students' Manuals.*

Chemistry:—Continued.

Trimble. Practical and Analytical Chemistry. A Course in Chemical Analysis, by Henry Trimble, Prof. of Analytical Chemistry in the Phila. College of Pharmacy. Illustrated. Second Edition. 8vo. Cloth, 1.50

Tidy. Modern Chemistry. 2d Ed. Cloth, 5.50

Leffmann's Compend of Chemistry. Inorganic and Organic. Including Urinary Analysis and the Sanitary Examination of Water. New Edition. Cloth, 1.00; Interleaved for Notes, 1.25

Müter. Practical and Analytical Chemistry. Second Edition. Revised and Illustrated. Cloth, 2.00

Holland. The Urine, Common Poisons, and Milk Analysis, Chemical and Microscopical. For Laboratory Use. 3d Edition, Enlarged. Illustrated. *In Press.*

Van Nüys. Urine Analysis. Illus. Cloth, 2.00

Wolff's Applied Medical Chemistry. By Lawrence Wolff, M.D., Demonstrator of Chemistry in Jefferson Medical College, Philadelphia. Cloth, 1.00

CHILDREN.

Goodhart and Starr. The Diseases of Children. A Manual for Students and Physicians. By J. F. Goodhart, M.D., Physician to the Evelina Hospital for Children; Assistant Physician to Guy's Hospital, London. American Edition, Revised and Edited by Louis Starr, M.D., Clinical Professor of Diseases of Children in the Hospital of the University of Pennsylvania; Physician to the Children's Hospital, Philadelphia. Containing many new Prescriptions, a List of over 50 Formulæ, conforming to the U. S. Pharmacopœia, and Directions for making Artificial Human Milk, for the Artificial Digestion of Milk, etc.
Cloth, 3.00; Leather, 3.50

Day. On Children. A Practical and Systematic Treatise. Second Edition. 8vo. 752 pages. Cloth, 3.00; Leather, 4.00

Meigs and Pepper. The Diseases of Children. Seventh Edition. 8vo. Cloth, 5.00; Leather, 6.00

Starr. Diseases of the Digestive Organs in Infancy and Childhood. With chapters on the Investigation of Disease, and on the General Management of Children. By Louis Starr, M.D., Clinical Professor of Diseases of Children in the University of Pennsylvania; with a section on Feeding, including special Diet Lists, etc. Illus. Cloth, 2.50

☞ *See pages 15 and 16 for list of ? Quiz-Compends ?*

DENTISTRY.

Fillebrown. Operative Dentistry. 330 Illustrations. *Just Ready.* Cloth, 2.50

Flagg's Plastics and Plastic Filling. 3d Ed. *Preparing.*

Gorgas. Dental Medicine. A Manual of Materia Medica and Therapeutics. Third Edition. Cloth, 3.25

Harris. Principles and Practice of Dentistry. Including Anatomy, Physiology, Pathology, Therapeutics, Dental Surgery and Mechanism. Twelfth Edition. Revised and enlarged by Professor Gorgas. 1028 Illustrations. Cloth, 7.00; Leather, 8.00

Richardson's Mechanical Dentistry. Fifth Edition. 569 Illustrations. 8vo. Cloth, 4.50; Leather, 5.50

Stocken's Dental Materia Medica. Third Edition. Cloth, 2.50

Taft's Operative Dentistry. Dental Students and Practitioners. Fourth Edition. 100 Illustrations. Cloth, 4.25; Leather, 5.00

Talbot. Irregularities of the Teeth, and their Treatment. Illustrated. 8vo. Cloth, 2.00

Tomes' Dental Anatomy. Third Ed. 191 Illus. *Preparing.*

Tomes' Dental Surgery. 3d Edition. Revised. 292 Illus. 772 Pages. Cloth, 5.00

DICTIONARIES.

Cleaveland's Pocket Medical Lexicon. Thirty-first Edition. Giving correct Pronunciation and Definition of Terms used in Medicine and the Collateral Sciences. Very small pocket size. Cloth, red edges .75; pocket-book style, 1.00

Longley's Pocket Dictionary. The Student's Medical Lexicon, giving Definition and Pronunciation of all Terms used in Medicine, with an Appendix giving Poisons and Their Antidotes, Abbreviations used in Prescriptions, Metric Scale of Doses, etc. 24mo. Cloth, 1.00; pocket-book style, 1.25

EYE.

Arlt. Diseases of the Eye. Including those of the Conjunctiva, Cornea, Sclerotic, Iris and Ciliary Body. By Prof. Von Arlt. Translated by Dr. Lyman Ware. Illus. 8vo. Cloth, 2.50

Hartridge on Refraction. 3d Ed. Cloth, 2.00

Macnamara. Diseases of the Eye. 4th Edition. Revised. Colored Plates and Wood Cuts and Test Types. Cloth, 4.00

Meyer. Diseases of the Eye. A complete Manual for Students and Physicians. 270 Illustrations and two Colored Plates. 8vo. Cloth, 4.50; Leather, 5.50

Fox and Gould. Compend of Diseases of the Eye and Refraction. 2d Ed. Enlarged. 71 Illus. 39 Formulæ. Cloth, 1.00; Interleaved for Notes, 1.25

☞ *See pages 2 to 5 for list of Students' Manuals.*

ELECTRICITY.
Mason's Compend of Medical and Surgical Electricity.
With numerous Illustrations. 12mo. Cloth, 1.00

HYGIENE.
Parkes' Practical Hygiene. Seventh Edition, enlarged. Illustrated. 8vo. Cloth, 4.50
Wilson's Handbook of Hygiene and Sanitary Science. Sixth Edition. Revised and Illustrated. Cloth, 2.75

MATERIA MEDICA AND THERAPEUTICS.
Potter's Compend of Materia Medica, Therapeutics and Prescription Writing. Fifth Edition, revised and improved.
Cloth, 1.00; Interleaved for Notes, 1.25
Biddle's Materia Medica. Eleventh Edition. By the late John B. Biddle, M.D., Professor of Materia Medica in Jefferson Medical College, Philadelphia. Thoroughly revised, and in many parts rewritten, by his son, Clement Biddle, M.D., Assistant Surgeon, U. S. Navy, assisted by Henry Morris, M.D., Demonstrator of Obstetrics in Jefferson Medical College. 8vo., illustrated. Cloth, 4.25; Leather, 5.00
Headland's Action of Medicines. 9th Ed. 8vo. Cloth, 3.00
Potter. Materia Medica, Pharmacy and Therapeutics. Including Action of Medicines, Special Therapeutics, Pharmacology, etc. *Page 4.* Cloth, 3.00; Leather, 3.50
Starr, Walker and Powell. Synopsis of Physiological Action of Medicines, based upon Prof. H. C. Wood's "Materia Medica and Therapeutics." 3d Ed. Enlarged. Cloth, .75
Waring. Therapeutics. With an Index of Diseases and an Index of Remedies. A Practical Manual. Fourth Edition. Revised and Enlarged. Cloth, 3.00; Leather, 3.50

MEDICAL JURISPRUDENCE.
Reese. A Text-book of Medical Jurisprudence and Toxicology. By John J. Reese, M.D., Professor of Medical Jurisprudence and Toxicology in the Medical Department of the University of Pennsylvania; President of the Medical Jurisprudence Society of Philadelphia; Physician to St. Joseph's Hospital; Corresponding Member of The New York Medico-legal Society. 2d Edition. Cloth, 3.00; Leather, 3.50
Woodman and Tidy's Medical Jurisprudence and Toxicology. Chromo-Lithographic Plates and 116 Wood engravings.
Cloth, 7.50; Leather, 8.50

☞ *See pages 15 and 16 for list of ? Quiz-Compends ?*

MISCELLANEOUS.

Allingham. Diseases of the Rectum. Fourth Edition. Illustrated. 8vo. Paper covers, .75; Cloth, 1.25

Beale. Slight Ailments. Their Nature and Treatment. Illustrated. 8vo. Paper cover, .75; Cloth, 1.25

Domville on Nursing. 6th Edition. Cloth, .75

Fothergill. Diseases of the Heart, and Their Treatment. Second Edition. 8vo. Cloth, 3.50

Gowers. Diseases of the Nervous System. 341 Illustrations. Cloth, 6.50; Leather, 7.50

Mann's Manual of Psychological Medicine, and Allied Nervous Diseases. Their Diagnosis, Pathology and Treatment, and their Medico-Legal Aspects. Illus. Cloth, 5.00; Leather, 6.00

Tanner. Memoranda of Poisons. Their Antidotes and Tests. Sixth Edition. Revised by Henry Leffmann, M.D. Cloth, .75

OBSTETRICS AND GYNÆCOLOGY.

Byford. Diseases of Women. The Practice of Medicine and Surgery, as applied to the Diseases and Accidents Incident to Women. By W. H. Byford, A.M., M.D., Professor of Gynæcology in Rush Medical College and of Obstetrics in the Woman's Medical College, etc., and Henry T. Byford, M.D., Surgeon to the Woman's Hospital of Chicago; Gynæcologist to St. Luke's Hospital, etc. Fourth Edition. Revised, Rewritten and Enlarged. With 306 Illustrations, over 100 of which are original. Octavo. 832 pages. Cloth, 5.00; Leather, 6.00

Cazeaux and Tarnier's Midwifery. With Appendix, by Mundé. The Theory and Practice of Obstetrics; including the Diseases of Pregnancy and Parturition, Obstetrical Operations, etc. By P. Cazeaux. Remodeled and rearranged, with revisions and additions, by S. Tarnier, M.D., Professor of Obstetrics and Diseases of Women and Children in the Faculty of Medicine of Paris. Eighth American, from the Eighth French and First Italian Edition. Edited by Robert J. Hess, M.D., Physician to the Northern Dispensary, Philadelphia, with an appendix by Paul F. Mundé, M.D., Professor of Gynæcology at the N. Y. Polyclinic. Illustrated by Chromo-Lithographs, Lithographs, and other Full-page Plates, seven of which are beautifully colored, and numerous Wood Engravings. *Students' Edition.* One Vol., 8vo. Cloth, 5.00; Leather, 6.00

Lewers' Diseases of Women. A Practical Text-Book. 139 Illustrations. Cloth, 2.25

Parvin's Winckel's Diseases of Women. Edited by Prof. Theophilus Parvin, Jefferson Medical College, Philadelphia. 117 Illustrations. *See page 3.* Cloth, 3.00; Leather, 3.50

Morris. Compend of Gynæcology. Illustrated. *In Press.*

☞ *See pages 2 to 5 for list of New Manuals.*

STUDENTS' TEXT-BOOKS AND MANUALS. 11

Obstetrics and Gynæcology:—Continued.

Winckel's Obstetrics. A Text-book on Midwifery, including the Diseases of Childbed. By Dr. F. Winckel, Professor of Gynæcology, and Director of the Royal University Clinic for Women, in Munich. Authorized Translation, by J. Clifton Edgar, M.D., Lecturer on Obstetrics, University Medical College, New York, with nearly 200 handsome illustrations, the majority of which are original with this work. Octavo. *In press.*

Landis' Compend of Obstetrics. Illustrated. 4th edition, enlarged. Cloth, 1.00; Interleaved for Notes, 1.25

Galabin's Midwifery. A New Manual for Students. By A. Lewis Galabin, M.D., F.R.C.P., Obstetric Physician to Guy's Hospital, London, and Professor of Obstetrics in the same Institution. 227 Illustrations. *See page 3.* Cloth, 3.00; Leather, 3.50

Glisan's Modern Midwifery. 2d Edition. Cloth, 3.00

Rigby's Obstetric Memoranda. By Alfred Meadows, M.D. 4th Edition. Cloth, .50

Meadows' Manual of Midwifery. Including the Signs and Symptoms of Pregnancy, Obstetric Operations, Diseases of the Puerperal State, etc. 145 Illustrations. 494 pages. Cloth, 2.00

Swayne's Obstetric Aphorisms. For the use of Students commencing Midwifery Practice. 8th Ed. 12mo. Cloth, 1.25

PATHOLOGY. HISTOLOGY. BIOLOGY.

Bowlby. Surgical Pathology and Morbid Anatomy, for Students. 135 Illustrations. 12mo. Cloth, 2.00

Davis' Elementary Biology. Illustrated. Cloth, 4.00

Rindfleisch's General Pathology. By Prof. Edward Rindfleisch. Translated by Wm. H. Mercur, M.D. Edited by James Tyson, M.D., Professor of Clinical Medicine in the University of Pennsylvania. 12mo. Cloth, 2.00

Gilliam's Essentials of Pathology. A Handbook for Students. 47 Illustrations. 12mo. Cloth, 2.00

*** The object of this book is to unfold to the beginner the fundamentals of pathology in a plain, practical way, and by bringing them within easy comprehension to increase his interest in the study of the subject.

Gibbes' Practical Histology and Pathology. Third Edition. Enlarged. 12mo. Cloth, 1.75

Virchow's Post-Mortem Examinations. 2d Ed. Cloth, 1.00

PHYSICAL DIAGNOSIS.

Bruen's Physical Diagnosis of the Heart and Lungs. By Dr. Edward T. Bruen, Assistant Professor of Clinical Medicine in the University of Pennsylvania. Second Edition, revised. With new Illustrations. 12mo. Cloth, 1.50

☞ *See pages 15 and 16 for list of ? Quiz-Compends ?*

PHYSIOLOGY.

Yeo's Physiology. Third Edition. The most Popular Students' Book. By Gerald F. Yeo, M.D., F.R.C.S., Professor of Physiology in King's College, London. Small Octavo. 758 pages. 321 carefully printed Illustrations. With a Full Glossary and Index. *See Page 3.* Cloth, 3.00; Leather, 3.50

Brubaker's Compend of Physiology. Illustrated. Fourth Edition. Cloth, 1.00; Interleaved for Notes, 1.25

Stirling. Practical Physiology, including Chemical and Experimental Physiology. 142 Illustrations. Cloth, 2.25

Kirke's Physiology. New 12th Ed. Thoroughly Revised and Enlarged. 502 Illustrations. Cloth, 4.00; Leather, 5.00

Landois' Human Physiology. Including Histology and Microscopical Anatomy, and with special reference to Practical Medicine. Third Edition. Translated and Edited by Prof. Stirling. 692 Illustrations. Cloth, 6.50; Leather, 7.50

"With this Text-book at his command, no student could fail in his examination."—*Lancet.*

Sanderson's Physiological Laboratory. Being Practical Exercises for the Student. 350 Illustrations. 8vo. Cloth, 5.00

Tyson's Cell Doctrine. Its History and Present State. Illustrated. Second Edition. Cloth, 2.00

PRACTICE.

Roberts' Practice. New Revised Edition. A Handbook of the Theory and Practice of Medicine. By Frederick T. Roberts, M.D.; M.R.C.P., Professor of Clinical Medicine and Therapeutics in University College Hospital, London. Seventh Edition. Octavo. Cloth, 5.50; Sheep, 6.50

Hughes. Compend of the Practice of Medicine. 3d Ed. Two parts, each, Cloth, 1.00; Interleaved for Notes, 1.25

PART I.—Continued, Eruptive and Periodical Fevers, Diseases of the Stomach, Intestines, Peritoneum, Biliary Passages, Liver, Kidneys, etc., and General Diseases, etc.

PART II.—Diseases of the Respiratory System, Circulatory System and Nervous System; Diseases of the Blood, etc.

Tanner's Index of Diseases, and Their Treatment. Cloth, 3.00

"This work has won for itself a reputation. . . . It is, in truth, what its Title indicates."—*N. Y. Medical Record.*

PRESCRIPTION BOOKS.

Wythe's Dose and Symptom Book. Containing the Doses and Uses of all the principal Articles of the Materia Medica, etc. Seventeenth Edition. Completely Revised and Rewritten. *Just Ready.* 32mo. Cloth, 1.00; Pocket-book style, 1.25

Pereira's Physician's Prescription Book. Containing Lists of Terms, Phrases, Contractions and Abbreviations used in Prescriptions, Explanatory Notes, Grammatical Construction of Prescriptions, etc., etc. By Professor Jonathan Pereira, M.D. Sixteenth Edition. 32mo. Cloth, 1.00; Pocket-book style, 1.25

☞ *See pages 2 to 5 for list of New Manuals.*

PHARMACY.

Stewart's Compend of Pharmacy. Based upon Remington's Text-Book of Pharmacy. Second Edition, Revised.
Cloth, 1.00 ; Interleaved for Notes, 1.25

SKIN DISEASES.

Anderson, (McCall) Skin Diseases. A complete Text-Book, with Colored Plates and numerous Wood Engravings. 8vo. *Just Ready.* Cloth, 4.50 ; Leather, 5.50

" We welcome Dr. Anderson's work not only as a friend, but as a benefactor to the profession, because the author has stricken off mediæval shackles of insuperable nomenclature and made crooked ways straight in the diagnosis and treatment of this hitherto but little understood class of diseases. The chapter on Eczema is alone worth the price of the book."—*Nashville Medical News.*

" Worthy its distinguished author in every respect; a work whose practical value commends it not only to the practitioner and student of medicine, but also to the dermatologist."—*James Nevens Hyde*, M.D., *Prof. of Skin and Venereal Diseases, Rush Medical College, Chicago.*

Van Harlingen on Skin Diseases. A Handbook of the Diseases of the Skin, their Diagnosis and Treatment (arranged alphabetically). By Arthur Van Harlingen, M.D., Clinical Lecturer on Dermatology, Jefferson Medical College ; Prof. of Diseases of the Skin in the Philadelphia Polyclinic. 2d Edition. Enlarged. With colored and other plates and illustrations. 12mo. Cloth, 2.50

Bulkley. The Skin in Health and Disease. By L. Duncan Bulkley, Physician to the N. Y. Hospital. Illus. Cloth, .50

SURGERY.

Jacobson. Operations in Surgery. A Systematic Handbook for Physicians, Students and Hospital Surgeons. By W. H. A. Jacobson, B.A., Oxon. F.R.C.S. Eng.; Ass't Surgeon Guy's Hospital ; Surgeon at Royal Hospital for Children and Women, etc. With 199 finely printed illustrations. 1006 pages. 8vo.
Cloth, $5.00 ; Leather, $6.00

Heath's Minor Surgery, and Bandaging. Eighth Edition. 142 Illustrations. 60 Formulæ and Diet Lists. Cloth, 2.00

Horwitz's Compend of Surgery, including Minor Surgery, Amputations, Fractures, Dislocations, Surgical Diseases, and the Latest Antiseptic Rules, etc., with Differential Diagnosis and Treatment. By ORVILLE HORWITZ, B.S., M.D., Demonstrator of Anatomy, Jefferson Medical College ; Chief, Out-Patient Surgical Department, Jefferson Medical College Hospital. 3d edition. Very much Enlarged and Rearranged. 91 Illustrations and 77 Formulæ. 12mo. *No. 9 ? Quiz-Compend ? Series.*
Cloth, 1.00 ; Interleaved for the addition of Notes, 1.25.

Pye's Surgical Handicraft. A Manual of Surgical Manipulations, Minor Surgery, Bandaging, Dressing, etc., etc. With special chapters on Aural Surgery, Extraction of Teeth, Anæsthetics, etc. 208 Illustrations. 8vo. Cloth, 5.00

Swain's Surgical Emergencies. New Edition. Illus. Clo., 1.50

☞ *See pages 15 and 16 for list of ? Quiz-Compends ?*

STUDENTS' TEXT-BOOKS AND MANUALS.

Surgery:—Continued.

Walsham. Manual of Practical Surgery. For Students and Physicians. By WM. J. WALSHAM, M.D., F.R.C.S., Asst. Surg. to, and Dem. of Practical Surg. in, St. Bartholomew's Hospital, Surgeon to Metropolitan Free Hospital, London. With 236 Engravings. *See Page 2.* Cloth, 3.00; Leather, 3.50

THROAT.

Mackenzie. Diseases of the Œsophagus, Nose and Naso-Pharynx. By Sir Morell Mackenzie, M.D., Senior Physician to the Hospital for Diseases of the Chest and Throat; Lecturer on Diseases of the Throat at the London Hospital, etc., with Formulæ and 93 Illustrations. Being Vol. II, complete in itself, of Dr. Mackenzie's text-book on the Throat and Nose.
Cloth, 3.00; Leather, 4.00

"It is both practical and learned; abundantly and well illustrated; its descriptions of disease are graphic and the diagnosis the best we have anywhere seen."—*Philadelphia Medical Times.*

Cohen. The Throat and Voice. Illustrated. Cloth, .50
James. Sore Throat. Its Nature, Varieties and Treatment. 12mo. Illustrated. Paper cover, .75; Cloth, 1.25

URINE, URINARY ORGANS, ETC.

Acton. The Reproductive Organs. In Childhood, Youth, Adult Life and Old Age. Sixth Edition. Cloth, 2.00
Beale. Urinary and Renal Diseases and Calculous Disorders. Hints on Diagnosis and Treatment. 12mo. Cloth, 1.75
Holland. The Urine, and Common Poisons. Chemical and Microscopical, for Laboratory Use. Illustrated. 2d Edition, Cloth, .75
Ralfe. Kidney Diseases and Urinary Derangements. 42 Illustrations. 12mo. 572 pages. Cloth, 2.75
Legg. On the Urine. A Practical Guide. 6th Ed. Cloth, .75
Marshall and Smith. On the Urine. The Chemical Analysis of the Urine. By John Marshall, M.D., Chemical Laboratory, Univ. of Penna; and Prof. E. F. Smith, PH.D. Col. Plates. Cloth, 1.00
Thompson. Diseases of the Urinary Organs. Eighth London Edition. Illustrated. Cloth, 3.50
Tyson. On the Urine. A Practical Guide to the Examination of Urine. With Colored Plates and Wood Engravings. 6th Ed. Enlarged. 12mo. Cloth, 1.50
—— Bright's Disease and Diabetes. Illus. Cloth, 3.50
Van Nüys, Urine Analysis. Illus. Cloth, 2.00

VENEREAL DISEASES.

Hill and Cooper. Student's Manual of Venereal Diseases, with Formulæ. Fourth Edition. 12mo. Cloth, 1.00
Durkee. On Gonorrhœa and Syphilis. Illus. Cloth, 3.50

☞ *See pages 15 and 16 for list of ? Quiz-Compends ?*

NEW AND REVISED EDITIONS.

?QUIZ-COMPENDS?

The Best Compends for Students' Use in the Quiz Class, and when Preparing for Examinations.

Compiled in accordance with the latest teachings of prominent lecturers and the most popular Text-books.

They form a most complete, practical and exhaustive set of manuals, containing information nowhere else collected in such a condensed, practical shape. Thoroughly up to the times in every respect, containing many new prescriptions and formulæ, and over two hundred and thirty illustrations, many of which have been drawn and engraved specially for this series. The authors have had large experience as quiz-masters and attachés of colleges, with exceptional opportunities for noting the most recent advances and methods. The arrangement of the subjects, illustrations, types, etc., are all of the most approved form, and the size of the books is such that they may be easily carried in the pocket. They are constantly being revised, so as to include the latest and best teachings, and can be used by students of any college of medicine, dentistry or pharmacy.

Cloth, each $1.00. Interleaved for Notes, $1.25.

No. 1. **HUMAN ANATOMY**, "Based upon Gray." Fourth Edition, including Visceral Anatomy, formerly published separately. Over 100 Illustrations. By SAMUEL O. L. POTTER, M.A., M.D., late A. A. Surgeon U. S. Army. Professor of Practice, Cooper Medical College, San Francisco.

Nos. 2 and 3. **PRACTICE OF MEDICINE.** Third Edition. By DANIEL E. HUGHES, M.D., Demonstrator of Clinical Medicine in Jefferson Medical College, Philadelphia. In two parts.

PART I.—Continued, Eruptive and Periodical Fevers, Diseases of the Stomach, Intestines, Peritoneum, Biliary Passages, Liver, Kidneys, etc. (including Tests for Urine), General Diseases, etc.

PART II.—Diseases of the Respiratory System (including Physical Diagnosis), Circulatory System and Nervous System; Diseases of the Blood, etc.

⁂ These little books can be regarded as a full set of notes upon the Practice of Medicine, containing the Synonyms, Definitions, Causes, Symptoms, Prognosis, Diagnosis, Treatment, etc., of each disease, and including a number of prescriptions hitherto unpublished.

(OVER.)

BLAKISTON'S ? QUIZ-COMPENDS ?
Continued.

Bound in Cloth, $1.00. Interleaved, for Notes, $1.25

No. 4. PHYSIOLOGY, including Embryology. Fourth Edition. By ALBERT P. BRUBAKER, M.D., Prof. of Physiology, Penn'a College of Dental Surgery; Demonstrator of Physiology in Jefferson Medical College, Philadelphia. Revised, Enlarged and Illustrated.

No. 5. OBSTETRICS. Illustrated. Fourth Edition. By HENRY G. LANDIS, M.D., Prof. of Obstetrics and Diseases of Women, in Starling Medical College, Columbus, O. Revised Edition. New Illustrations.

No. 6. MATERIA MEDICA, THERAPEUTICS AND PRESCRIPTION WRITING. Fifth Revised Edition. With especial Reference to the Physiological Action of Drugs, and a complete article on Prescription Writing. Based on the Last Revision of the U. S. Pharmacopœia, and including many unofficinal remedies. By SAMUEL O. L. POTTER, M.A., M.D., late A. A. Surg. U. S. Army; Prof. of Practice, Cooper Medical College, San Francisco. Improved and Enlarged, with Index.

No. 7. GYNÆCOLOGY. A Compend of Diseases of Women. By HENRY MORRIS, M.D., Demonstrator of Obstetrics, Jefferson Medical College, Philadelphia. *In Press.*

No. 8. DISEASES OF THE EYE AND REFRACTION, including Treatment and Surgery. By L. WEBSTER FOX, M.D., Chief Clinical Assistant Ophthalmological Dept., Jefferson Medical College, etc., and GEO. M. GOULD, M.D. 71 Illustrations, 39 Formulæ. Second Enlarged and Improved Edition. Index.

No. 9. SURGERY. Illustrated. Third Edition. Including Fractures, Wounds, Dislocations, Sprains, Amputations and other operations; Inflammation, Suppuration, Ulcers, Syphilis, Tumors, Shock, etc. Diseases of the Spine, Ear, Bladder, Testicles, Anus, and other Surgical Diseases. By ORVILLE HORWITZ, A.M., M.D., Demonstrator of Anatomy, Jefferson Medical College. Revised and Enlarged. 77 Formulæ and 91 Illustrations.

No. 10. CHEMISTRY. Inorganic and Organic. For Medical and Dental Students. Including Urinary Analysis and Medical Chemistry. By HENRY LEFFMANN, M.D., Prof. of Chemistry in Penn'a College of Dental Surgery, Phila. A new Edition, Revised and Rewritten, with Index.

No. 11. PHARMACY. Based upon "Remington's Text-book of Pharmacy." By F. E. STEWART, M.D., PH.G., Quiz-Master at Philadelphia College of Pharmacy. Second Edition, Revised.

Bound in Cloth, $1. Interleaved, for the Addition of Notes, $1.25.

☞ *These books are constantly revised to keep up with the latest teachings and discoveries, so that they contain all the new methods and principles. No series of books are so complete in detail, concise in language, or so well printed and bound. Each one forms a complete set of notes upon the subject under consideration.*

www.ingramcontent.com/pod-product-compliance
Lightning Source LLC
Chambersburg PA
CBHW031933230426
43672CB00010B/1906